The Princeton Review®

Cracking the

AP®

ENVIRONMENTAL SCIENCE EXAM

2015 Edition

Angela Morrow, PhD, and Tim Ligget

PrincetonReview.com

PENGUIN RANDOM HOUSE

Random House, Inc. New York

The Princeton Review
24 Prime Parkway, Suite 201
Natick, MA 01760
E-mail: editorialsupport@review.com

Published in the United States by Random House, LLC, New York, and
simultaneously in Canada by Random House of Canada Limited, Toronto.

A Penguin Random House Company.

ISBN: 978-0-8041-2532-1
eBook ISBN: 978-0-8041-2533-8
ISSN: 1558-9684

AP and Advanced Placement Program are registered trademarks of the
College Board, which does not sponsor or endorse this product.

The Princeton Review is not affiliated with Princeton University.

Editor: Meave Shelton
Production Editor: Lee Elder
Production Artist: Keren Peysakh

Printed in the United States of America on partially recycled paper.

10 9 8 7 6 5 4 3 2 1

2015 Edition

Editorial
Rob Franek, Senior VP, Publisher
Casey Cornelius, VP Content Development
Mary Beth Garrick, Director of Production
Selena Coppock, Managing Editor
Calvin Cato, Editor
Colleen Day, Editor
Aaron Riccio, Editor
Meave Shelton, Editor

Random House Publishing Team
Tom Russell, Publisher
Alison Stoltzfus, Publishing Manager
Dawn Ryan, Associate Managing Editor
Ellen L. Reed, Production Manager
Erika Pepe, Associate Production Manager
Kristin Lindner, Production Supervisor
Andrea Lau, Designer

Acknowledgments

This book is written in honor of my son, Stephen, who has been willing to share his birthdays and vacations with APES for the past eight years. Thank you Charles, Samantha, and Stephen for being proud of having a scientist as a mother and for all of your encouragement and understanding. George, thank you for sustaining me, and being agreeable to eating out and living with a dirty house while I was involved in this project. I appreciate you taking me to the airport in the wee hours and most importantly picking me up again. Thanks also to Rachel Warren.

—Angela Morrow

I would first like to thank my wife Denise, daughters Deirdre and Sarah, and my son J. T. for their patience and understanding during the time I was working on this edition. Their encouragement made the effort so worthwhile. Secondly, I would like to thank the faculty of Gettysburg College. They nurtured my lifelong love of learning and the biological sciences. Finally, I give my thanks to all my friends at the AP readings. They helped with my professional development and renewed my love of Environmental Science every year.

—Tim Ligget

Special thanks to Eliz Markowitz for her revision of this year's edition.

Contents

Part I
Using This Book
to Improve
Your AP Score

- Preview: Your Knowledge, Your Expectations
- Your Guide to Using This Book
- How to Begin

PREVIEW: YOUR KNOWLEDGE, YOUR EXPECTATIONS

Your route to a high score on the AP Environmental Science Exam depends a lot on how you plan to use this book. Respond to the following questions:

1. Rate your level of confidence about your knowledge of the content tested by the AP Environmental Science Exam.

   A. Very confident—I know it all

   B. I'm pretty confident, but there are topics for which I could use help

   C. Not confident—I need quite a bit of support

   D. I'm not sure.

2. Circle your goal score for the AP Environmental Science Exam.

   5   4   3   2   1   I'm not sure yet

3. What do you expect to learn from this book? Circle all that apply to you.

   A. A general overview of the test and what to expect

   B. Strategies for how to approach the test

   C. The content tested by this exam

   D. I'm not sure yet

YOUR GUIDE TO USING THIS BOOK

This book is organized to provide as much—or as little—support as you need, so you can use this book in whatever way will be most helpful to improving your score on the AP Environmental Science Exam.

- The remainder of Part I will provide guidance on how to use this book and help you determine your strengths and weaknesses.

- Part II of this book will:
    - o provide information about the structure, scoring, and content of the AP Environmental Science Exam
    - o help you to make a study plan
    - o point you towards additional resources

- Part III of this book will explore the following strategies:
    - o how to attack multiple-choice questions
    - o how to approach free-response questions
    - o how to manage your time to maximize the number of points available to you

- Part IV of this book covers the content you need for the AP Environmental Science Exam.

- Part V of this book contains practice tests, answers and explanations, and scoring guides. You may choose the use some parts of this book over others, or you may work through the entire book. This will depend on your needs and how much time you have. Let's now examine how to make this determination.

HOW TO BEGIN

1. **Take a Test**

    Before you can decide how to use this book, you need to take a practice test. Doing so will give you insight into your strengths and weaknesses, and help you make an effective study plan. If you're feeling test-phobic, remind yourself that a practice test is a tool for diagnosing yourself—it's not how well you do that matters, but how you use information gleaned from your performance to guide your preparation.

    So, before you read further, take Practice Test 1 starting at page 285 of this book. Be sure to do so in one sitting, following the instructions that appear before the test.

2. **Check Your Answers**

    Using the answer key on page 304, count how many multiple-choice questions you got right and how many you missed. Don't worry about the explanations for now, and don't worry about why you missed questions. We'll get to that soon.

3. **Reflect on the Test**

    After you take your first test, respond to the following questions:
    *   How much time did you spend on the multiple-choice questions?
    *   How much time did you spend on each free-response question?
    *   How many multiple-choice questions did you miss?
    *   Do you feel you had the knowledge to address the subject matter of the free-response questions?
    *   Circle the content areas that were most challenging for you and draw a line through the ones in which you felt confident/did well.

    Geological eras
    Atmosphere
    Soil dynamics
    Population
    Pollution (specifically solid waste, acid rain, ozone depletion, and thermal)
    Water conservation
    Conservation & economic impact
    Interpreting charts and graphs
    Interpreting experimental data
    Analyzing experiments & experimental reasoning
    Ecosystems & biodiversity
    Biogeochemical cycles
    Global climate change
    Environmental regulation and international treaties
    Energy resources & consumption
    Energy science
    Mining

4. Read Part II of this Book and Complete the Self-Evaluation

As discussed in the Guide section above, Part II will provide information on how the test is structured and scored. It will also outline areas of content that are tested.

As you read Part II, re-evaluate your answers to the questions you just answered. At the end of Part II, you will revisit and refine your answers to the previous questions. You will then be able to make a study plan, based on your needs and time available, that will allow you to use this book most effectively.

5. Engage with Parts III and IV as Needed

Notice the word engage. You'll get more out of this book if you use it intentionally than if you read it passively, hoping for an improved score through osmosis.

Strategy chapters will help you think about your approach to the question types on this exam. Part III will open with a reminder to think about how you approach questions now, and then close with a reflection section asking you to think about how/whether you will change your approach in the future.

Content chapters are designed to provide a review of the content tested on the AP Environmental Science Exam, including the level of detail you need to know and how the content is tested. You will have the opportunity to assess your mastery of the content of each chapter through test-appropriate questions and a reflection section.

6. Take Practice Test 2 and Assess Your Performance

Once you feel you have developed the strategies you need and gained the knowledge you lacked, you should take Practice Test 2, which starts at page 323 of this book. You should do so in one sitting, following the instructions at the beginning of the test.

When you are done, check your answers to the multiple-choice sections. See if a teacher will read your responses to the free-response questions and provide feedback.

Once you have taken the test, reflect on what areas you still need to work on, and revisit the chapters in this book that address those deficiencies. Through this type of reflection and engagement, you will continue to improve.

7. Keep Working

As Part II will discuss, there are other resources available to you, including a wealth of information on AP Central. You can continue to explore areas where you can stand to improve and engage in those areas right up to the day of the test.

Part II
About the AP Environmental Science Exam

- The Structure of the AP Environmental Science Exam
- How the AP Environmental Science Exam Is Scored
- Overview of Content Topics
- How AP Exams Are Used
- Other Resources
- Designing Your Study Plan

THE STRUCTURE OF THE AP ENVIRONMENTAL SCIENCE EXAM

The AP Environmental Science Exam is three hours long. You have 90 minutes to answer 100 multiple-choice questions and 90 minutes to answer four free-response questions.

Part I: The Multiple-Choice Section

Multiple-choice questions come in two types. The first looks like this:

Questions 1–5 refer to the structure of the atmosphere.

> (A) Troposphere
> (B) Stratosphere
> (C) Thermosphere
> (D) Mesosphere
> (E) Stratopause

1. The layer that contains the earth's daily weather

2. Extends from 50 to 80 km in height

3. Contains the majority of the earth's ozone layer

4. The highest layer of the atmosphere (above 80 km)

5. The layer of the atmosphere heated by the IR radiation from the earth

This is basically a matching question that's dressed up to look more sophisticated. About the first 30 or so questions in the multiple-choice section will be in this form, and we'll talk about how you should tackle these questions in Chapter 1. We'll discuss the science underlying these questions later on, but in case you're interested—the answers are (A), (D), (B), (C), and (A), respectively.

The second type of multiple-choice question is more like the traditional multiple-choice questions you are used to seeing:

> 7. Salt intrusion into freshwater aquifers, beach erosion, and disruption of coastal fisheries all might occur as a result of
>
> (A) rising ocean levels as global warming proceeds
> (B) more solar ultraviolet radiation on the earth
> (C) more chlorofluorocarbons in the atmosphere
> (D) reduced rates of photosynthesis
> (E) using the oceans as a waste disposal area

There are about 70 such questions.

Part II: The Free-Response Section

Environmental Science is interdisciplinary. This means that it draws from several sciences (biology, chemistry, and physics) and the humanities (government, economics, and social studies). Your answers to free-response questions need to take a comprehensive view and draw from several disciplines. A free-response question may look like this:

3. According to the United States Energy Information Administration, the consumption of natural gas in the United States increases 8 percent per year. The United States receives its supplies of gas from a variety of international and domestic locations. Natural gas is used in the home, for industry, and for power generation.

   (a) Calculate the approximate number of years it will take to double the consumption of natural gas. Show all work.
   (b) Describe one method for the recovery and transportation of natural gas.
   (c) Describe two benefits to the environment that would occur by switching from coal to natural gas-fired electric power generation.
   (d) Some people advocate increasing the use of coal instead of natural gas for the production of electricity. Explain one argument that the proponents of coal might use to justify their position.

As you can see, for this multipart free-response question, your answers should not be one-dimensional; they will need to encompass many different subjects and areas of thought.

We'll talk more about how to go about writing your responses to these questions in Chapter 2, but for now, remember that you have only 90 minutes to answer all four questions. This translates to about 22 minutes per question, so any practice you can get before test day will be invaluable! Fortunately for you, we have put sample free-response answers in each chapter of this book to give you that practice.

Don't underestimate the difficulty of this exam! More people receive a score of 1 on the AP Environmental Science exam than any other AP Exam!

HOW THE AP ENVIRONMENTAL SCIENCE EXAM IS SCORED

What Will My Score Look Like, and What Will it Mean?

After taking the test in mid-May, you will receive your score sometime around the first week of July, right around the time when you've just started to forget about the entire experience. Your score will be a single number from 1 to 5. Here's what those numbers mean.

	May 2013 Score Results		
Score	What this score means	Approximate % of all test takers receiving this score	Will a student with this score receive credit?
5	Extremely Qualified	7.7	Yes
4	Well Qualified	23.3	Yes
3	Qualified	16.9	Maybe
2	Possibly Qualified	25.6	Very Rarely
1	Not Qualified	26.5	No

Those percentages are pretty intimidating, huh? Quite a few students get a 1 on this exam. Why is this? Well, it's probably because these students were not willing to put the time and energy into studying and reviewing the necessary topics.

However, by purchasing this book, you've already proven that you aren't one of those students. No one said this test would be easy, but it is definitely manageable.

How Much Will Each Section Count Toward My Final Score?

As we mentioned, a computer will grade the multiple-choice section of your exam. Your final grade will be made up of the two sections, which will be given different weights: 60 percent of the grade will come from the 100 multiple-choice questions, and the remaining 40 percent will come from the four free-response questions, with about 10 percent coming from each of your responses.

Your Multiple-Choice Score

For this section of the test, you will be scored only on the number of questions you answer correctly. That's right, only the ones you get right are counted! Since there is no longer any penalty for wrong answers, it is in your best interest to answer as many multiple choice questions as possible. Although random guessing is better than nothing, you should use Process of Elimination (POE) to narrow down the choices first. We'll get into the details of POE in Chapter 1. If you are running out of time, remember, you need to fill in all the bubbles before time is up. If you don't have time for POE, just choose any letter and move on.

Your Free-Response Score

Each AP free-response question is scored on a scale from 0 to 10, with 10 being the best score. The scores of all four free-response questions are added together to obtain your free-response score.

At the beginning of the free-response question grading, rubrics are formulated by the Chief Reader, a university professor who also attends development committee meetings, and other AP Environmental Science leaders. These rubrics are refined again and again to provide the best possible scoring rubric for each question. Leaders for each question will direct readers and reread tests to make sure that every test is graded accurately and fairly according to the rubric. Computer-generated statistics aid the Chief Reader and the other leaders in this quality assurance, and each reader (the readers are either high school AP Environmental Science teachers or university professors) is assigned to read only one free-response answer and becomes an expert on that question.

As you can imagine, reading a thousand or more test papers in eight days is an awesome task; therefore, anything you can do to make your free-response answers easier to read is appreciated! A well-organized and readable exam may receive a higher score because the reader can actually find the information he or she needs in order to award the earned points.

Each free-response question is worth 10 points, and there are several different ways to earn those 10 points. However, you can never earn all 10 points unless you answer all parts of the free-response question. So if the free-response question has 4 parts, you cannot score 10 points unless you answer all 4 parts of that question correctly!

We've said that the highest possible score on each question is a 10. Answers that receive a 10 demonstrate a clear and thorough knowledge of the material, and include a superb response to all parts of the question. If the answer requires a calculation, the student performs the calculation showing all work and labeling all units. If it includes placing points on a graph or graphing an answer, the graph is drawn correctly and the points are placed in the correct place, with all elements (including the x and y axes) labeled correctly.

Answers that receive a 9 also demonstrate a thorough knowledge of the material, but are docked a point due to sloppy work; the student might not have labeled a graph correctly or shown their work in doing a calculation, for example. Answers that score a 7 or 8 do not give a satisfactory response to one part of the question, while answers that score a 3, 4, or 5 give an unsatisfactory response for more than one part.

Sometimes a student may read a question and decide before attempting it that he or she doesn't know how to draw the graph or do the necessary calculations, so he or she skips that part. But that's the wrong thing to do! Always attempt all parts of a question; you might be surprised to find that it is easier than you think.

In all cases, even if you don't think you know the answer to one or more questions, you should thoroughly read the question, complete the rest of the free-response questions in the section and then come back to it.

To look at real free-response questions from years past, go to **https://apstudent. collegeboard.org/apcourse/ap-environmental-science/exam-practice.**

Your Final Score

Your final 1 to 5 score is a combination of your section scores. Remember that the multiple-choice section counts for 60 percent of the total and that the essays count for 40 percent. However, the bottom line is that both sections are very important, and you must concentrate on doing your best on both parts.

Keep in mind that even if you do not get college credit for this course, you will not have wasted your time. Research shows that just taking an AP course helps your college performance. Small consolation perhaps, but as we mentioned earlier, because you're making this effort to prepare properly, you'll most likely get due credit.

But How Do I Get Credit If (I Mean, When) I Score a 4 or 5?

While your score will be sent to schools, you should keep a copy of your score and take it when you register for classes at the college or university you attend.

Also, as this is one of the newer AP courses—and an interdisciplinary course—you should probably keep the syllabus and your laboratory notebook from your AP Environmental Science course. You may need to show it to college or university counselors in order to get college credit for a science laboratory course.

OVERVIEW OF CONTENT TOPICS

So what do you really need to know about environmental science in order to do well on the exam? The next several pages outline the topics that the College Board says should be covered in your course (and thus, on the exam). The percentages give you a rough idea of the balance of questions from each category on the test.

I. **Earth Systems and Resources (10–15 percent)**
    a. Earth Science Concepts
    - Geologic time scale
    - Plate tectonics
    - Earthquakes
    - Volcanism
    - Seasons
    - Solar intensity
    - Latitude

    b. The Atmosphere
    - Composition
    - Structure
    - Weather and climate
    - Atmospheric circulation and the Coriolis Effect
    - Atmosphere-ocean interactions
    - El Niño and Southern Oscillation (ENSO)

    c. Global Water Resources and Uses
    - Freshwater/saltwater
    - Ocean circulation
    - Agricultural, industrial, and domestic use
    - Surface and groundwater issues
    - Global challenges and conservation

    d. Soil and Soil Dynamics
    - Rock cycle
    - Formation
    - Composition
    - Physical and chemical properties
    - Main soil types
    - Erosion and other soil problems
    - Soil conservation

II. **The Living World (10–15 percent)**
    a. Ecosystem Structure
    - Biological populations and communities
    - Ecological niches
    - Interactions among species
    - Keystone species
    - Species diversity and edge effect
    - Major terrestrial and aquatic biomes

b. Energy Flow
- Photosynthesis and cellular respiration
- Food webs and trophic levels
- Ecological pyramids

c. Ecosystem Diversity
- Biodiversity
- Natural selection
- Evolution
- Ecosystem services

d. Natural Ecosystem Cycles
- Climate shifts
- Species movement
- Ecological succession

e. Natural Biogeochemical Cycles
- Carbon
- Nitrogen
- Phosphorus
- Sulfur
- Water
- Conservation of matter

III. Population (10–15 percent)
a. Population Biology Concepts
- Population ecology
- Carrying capacity
- Reproductive strategies
- Survivorship

1. Human population
- Historical population sizes
- Distribution
- Fertility rates
- Growth rates and doubling times
- Demographic transition
- Age-structure diagrams

2. Population size
- Strategies for sustainability
- Case studies
- National policies

3. Impacts of population growth
- Hunger
- Disease
- Economic effects
- Resource use
- Habitat destruction

IV. Land and Water Use (10–15 percent)
a. Agriculture
   1. Feeding a growing population
- Human nutritional requirements
- Types of agriculture
- Green Revolution
- Genetic engineering and crop production
- Deforestation
- Irrigation
- Sustainable agriculture

   2. Controlling pests
- Types of pesticides
- Costs and benefits of pesticide use
- Integrated pest management
- Relevant laws

b. Forestry
- Tree plantations
- Old growth forests
- Forest fires
- Forest management
- National forests

c. Rangelands
- Overgrazing
- Deforestation
- Desertification
- Rangeland management
- Federal rangelands

d. Other land uses
   1. Urban land development
- Planned development
- Suburban sprawl
- Urbanization

   2. Transportation infrastructure
- Federal highway system
- Canals and channels
- Roadless areas
- Ecosystem impacts

   3. Public and federal lands
- Management
- Wilderness areas
- National parks
- Wildlife refuges
- Forests
- Wetlands

4. Land conservation options
- Preservation
- Remediation
- Mitigation
- Restoration
- Sustainable Land-Use Strategies

e. Mining
- Mineral formation
- Extraction
- Global reserves
- Relevant laws and treaties

f. Fishing
- Fishing techniques
- Overfishing
- Aquaculture
- Relevant laws and treaties

g. Global Economics
- Globalization
- World Bank
- Tragedy of the Commons
- Relevant laws and treaties

V. **Energy Resources and Consumption (10–15 percent)**
a. Energy Concepts
- Energy forms
- Power
- Units
- Conversions
- Laws of Thermodynamics

b. Energy Consumption
1. History
- Industrial Revolution
- Exponential growth
- Energy crisis

2. Present global energy use

3. Future energy needs

c. Fossil Fuel Resources and Uses
- Formation of coal, oil, and natural gas
- Extraction and purification methods
- World reserves and global demand
- Synfuels
- Environmental advantages and disadvantages of sources

d. Nuclear Energy
- Nuclear fission process
- Nuclear fuel
- Electricity production
- Nuclear reactor types
- Environmental advantages and disadvantages
- Safety issues
- Radiation and human health
- Radioactive wastes
- Nuclear fusion

e. Hydroelectric Power
- Dams
- Flood control
- Salmon
- Silting
- Other impacts

f. Energy Conservation
- Energy efficiency
- CAFE standards
- Hybrid electric vehicles
- Mass transit

g. Renewable Energy
- Solar energy
- Solar electricity
- Hydrogen fuel cells
- Biomass
- Wind energy
- Small-scale hydroelectric
- Ocean waves and tidal energy
- Geothermal
- Environmental advantages/disadvantages

VI. Pollution (25–30 percent)
a. Polution types
1. Air pollution
- Primary and secondary sources
- Major air pollutants
- Measurement units
- Smog
- Acid deposition—causes and effects
- Heat islands and temperature inversions
- Indoor air pollution
- Remediation and reduction strategies
- Clean Air Act and other relevant laws

2. Noise pollution
- Sources
- Effects
- Control measures

3. Water pollution
- Types
- Sources, causes, and effects
- Cultural eutrophication
- Groundwater pollution
- Maintaining water quality
- Water purification
- Sewage treatment/septic systems
- Clean Water Act and other relevant laws

4. Solid Waste
- Types
- Disposal
- Reduction

b. Impacts on the Environment and Human Health
  1. Hazards to human health
  - Environmental risk analysis
  - Acute and chronic effects
  - Dose-response relationships
  - Air pollutants
  - Smoking and other risks

  2. Hazardous chemicals in the environment
  - Types of hazardous waste
  - Treatment/disposal of hazardous waste
  - Cleanup of contaminated sites
  - Biomagnification
  - Relevant laws

c. Economic Impacts
- Cost-benefit analysis
- Externalities
- Marginal costs
- Sustainability

VII. Global Change (10–15 percent)
a. Stratospheric Ozone
- Formation of stratospheric ozone
- Ultraviolet radiation
- Causes and effects of ozone depletion
- Strategies for reducing ozone depletion
- Relevant laws and treaties

b. Global Warming
- Greenhouse gases and the greenhouse effect
- Impacts and consequences of global warming
- Reducing climate change
- Relevant laws and treaties

c. Loss of Biodiversity
- Habitat loss
- Overuse
- Pollution
- Introduced species
- Endangered and extinct species
- Maintenance through conservation
- Relevant laws and treaties

Whew. Again, that's a lot. But in the content review of this book, we will go over all of the material—and only the material—that you really need to know to score high on the test.

HOW AP EXAMS ARE USED

Different colleges use AP exam scores in different ways, so it is important that you go to a particular college's web site to determine how it uses AP exam scores. The three items below represent the main ways in which AP exam scores can be used:

- **College Credit**. Some colleges will give you college credit if you score well on an AP exam. These credits count towards your graduation requirements, meaning that you can take fewer courses while in college. Given the cost of college, this could be quite a benefit, indeed.

- **Satisfy Requirements**. Some colleges will allow you to "place out" of certain requirements if you do well on an AP exam, even if they do not give you actual college credits. For example, you might not need to take an introductory-level course, or perhaps you might not need to take a class in a certain discipline at all.

- **Admissions Plus**. Even if your AP exam will not result in college credit or even allow you to place out of certain courses, most colleges will respect your decision to push yourself by taking an AP course or even an AP exam outside of a course. A high score on an AP exam shows mastery of more difficult content than is taught in many high school courses, and colleges may take that into account during the admissions process.

OTHER RESOURCES

There are many resources available to help you improve your score on the AP Environmental Science Exam, not the least of which are your **teachers**. If you are taking an AP class, you may be able to get extra attention from your teacher, such as obtaining feedback on your free-response essays. If you are not in an AP course, reach out to a teacher who teaches science, and ask if the teacher will review your essays or otherwise help you with content.

Another wonderful resource is **AP Central**, the official site of the AP exams. The scope of the information at this site is quite broad and includes:

- A Course Description, which provides details on what content is covered and sample questions

- Official released AP Environmental Science Exams, available for purchase at the College Board store

- One free 1998 AP Environmental Science Released Exam

- Practice multiple-choice questions

- Free-response prompts from previous years

- Lab activities and resources

The AP Central home page address is: **http://apcentral.collegeboard.com/apc/ Controller.jpf**

The AP Environmental Science Course home page address is: **https://apstudent. collegeboard.org/apcourse/ap-environmental-science?envsci**

Finally, **The Princeton Review** offers tutoring for the AP Environmental Science Exam. Our expert instructors can help you refine your strategic approach and add to your content knowledge. For more information, call 1-800-2REVIEW.

DESIGNING YOUR STUDY PLAN

In Part I you identified some areas of potential improvement. Let's now delve further into your performance on Practice Test 1, with the goal of developing a study plan appropriate to your needs and time commitment.

Read the answers and explanations associated with the multiple-choice questions (starting at page 303). After you have done so, respond to the following questions:

- Review the Overview of Content Topics on pages 13–19, and, next to each one, indicate your rank of the topic as follows: "1" means "I need a lot of work on this," "2" means "I need to beef up my knowledge," and "3" means "I know this topic well."

- How many days/weeks/months away is your AP Environmental Science Exam?

- What time of day is your best, most focused study time?

- How much time per day/week/month will you devote to preparing for your AP Environmental Science Exam?

- When will you do this preparation? (Be as specific as possible: Mondays & Wednesdays from 3 to 4 p.m., for example)

- Based on the answers above, will you focus on strategy (Part III) or content (Part IV) or both?

- What are your overall goals in using this book?

Part III
Test-Taking
Strategies
for the AP
Environmental
Science Exam

PREVIEW

Review your responses to the questions on page 4 of Part I and then respond to the following questions:

- How many multiple-choice questions did you miss even though you knew the answer?

- On how many multiple-choice questions did you guess blindly?

- How many multiple-choice questions did you miss after eliminating some answers and guessing based on the remaining answers?

- Did you find any of the free-response questions easier/harder than the others—and, if so, why?

HOW TO USE THE CHAPTERS IN THIS PART

Think about what you are doing now before you read the following Strategy chapters. As you read and engage in the directed practice, be sure to appreciate the ways you can change your approach. At the end of Part III, you will have the opportunity to reflect on how you will change your approach.

Chapter 1
How to
Approach
Multiple-Choice
Questions

THE PRINCETON REVIEW APPROACH

There are basically two ways to prepare for the AP Environmental Science Exam.

- Know absolutely everything about everything. Bad idea.

- Review only what you need to know, and tackle the test strategically. Good idea.

This is The Princeton Review's way—and the best way—to improve your score.

Rather than trying to teach you everything there is to know about environmental science, we at The Princeton Review focus on test-taking strategies. Naturally, we'll review some hard science as well. But rather than cluttering your brain, we'll look only at the environmental science you need to know for the test, explaining and highlighting key concepts along the way.

First, we'll give you some simple, straightforward strategies for tackling multiple-choice questions and for writing free-response answers. Let's now take a closer look at how to approach the multiple-choice section.

The Two-Pass System

The AP Environmental Science Exam covers a broad range of topics. There's no way, even with our extensive review, that you will know everything about every topic in environmental science. So, what should you do?

Adopt a two-pass system. The two-pass system entails going through the test and answering the easy questions first. Save the more time-consuming questions for later. (Don't worry—you'll have time to do them later!) First, read the question and decide if it a "now" or "later" question. If you decide this is a "now" question, answer it in the test booklet. If it is a "later" question, come back to it. Once you have finished all the "now" questions on a double page, transfer the answers to your bubble sheet. Flip the page and repeat the process.

Once you've finished all the "now" questions, move on to the "later" questions. Start with the easier questions first. These are the ones that require calculations or that require you to eliminate the answer choices (in essence, the correct answer does not jump out at you immediately). Transfer your answers to your bubble sheet as soon as you answer these "later" questions.

Watch Out for Those Bubbles!

Because you're skipping problems, you need to keep careful track of the bubbles on your answer sheet. One way to accomplish this is by answering all the questions on a page and then transferring your choices to the answer sheet. If you prefer to enter them one by one, make sure you double-check the number beside the ovals before filling them in. We'd hate to see you lose points because you forgot to skip a bubble!

The two-pass system allows you to pick up easy points from the start!

Process of Elimination (POE)

On most tests, you need to know your material backward and forward in order to get the right answer. In other words, if you don't know the answer beforehand, you probably won't answer the question correctly. This is particularly true of fill-in-the-blank and essay questions. We're taught to think that the only way to get a question right is by knowing the answer. However, that's not the case on Section I of the AP Environmental Science Exam. You can get a perfect score on this portion of the test without knowing a single right answer—provided you know all the wrong answers!

What are we talking about? This is perhaps the most important technique to use on the multiple-choice section of the exam. Let's take a look at an example.

41. The long-term storage of phosphorus and sulfur
    occurs in which of the following?

    (A) Bacteria
    (B) Rocks
    (C) Water
    (D) Plants
    (E) Atmosphere

Now if this were a fill-in-the-blank-style question, you might be in a heap of trouble. But let's take a look at what we've got. You see the elements phosphorus and sulfur in the question, which leads you to conclude that we're talking about elements. Right away, you can probably remember that these aren't normally components of water, so you can eliminate (C). Also, plants don't live a long time, so sulfur and phosphorus can't be stored for the long-term in plants, right? Get rid of (D). The same goes for bacteria, so lose (A). You're left with (B) and (E). If you know that neither of these elements is a significant component of the atmosphere, then you can get rid of (E) and see that the best answer is (B), but even if you don't, you have a fifty-fifty chance of guessing the correct answer at this point.

We think we've illustrated our point: Process of Elimination is the best way to approach the multiple-choice questions. Even when you don't know the answer right off the bat, you'll surely know that two or three of the answer choices are not correct. What then?

It is often easier to identify a wrong answer than a right answer. Use POE to get rid of bad answers!

Aggressive Guessing

As mentioned earlier, you are scored only on the number of questions you get right, so we know guessing can't hurt you. But can it help you? It sure can. Let's say you guess on five questions; odds are you'll get one right. So you've already increased your score by one point. Now, let's add POE into the equation. If you can eliminate as many as two answer choices from each question, your chances of getting them right increase, and so does your overall score. Remember: Don't leave any bubbles blank on test day!

Word Associations

Another way to rack up the points on the AP Environmental Science Exam is by using word associations in tandem with your POE skills. Make sure that you memorize all of the words in the Hit Parade, which is Chapter 12 of this book. Know them backward and forward. As you learn them, make sure you group them by "association," since ETS is bound to ask about them on the AP Environmental Science Exam. What do we mean by "word associations"?

Let's take the example of air pollution. You'll soon see from our review, and possibly your course study, that there are several compounds associated with various types of air pollution; for example, ozone, VOC, and nitrogen oxides are all words and terms that are associated with air pollution. Now take a look below at a typical question about pollution.

> 2. All of the following are important in smog production EXCEPT
>
> (A) photochemical reactions
> (B) stratospheric ozone
> (C) tropospheric ozone
> (D) volatile organic compounds
> (E) nitrogen oxides

This might seem like a difficult question, but let's think about the associations we just discussed. The question asks us about smog. Answer choices (C), (D), and (E) are all terms that we've associated with air pollution. Therefore, we can eliminate them. Maybe you're unsure about whether or not photochemical reactions are part of air pollution, but since you know for sure that stratospheric ozone has nothing to do with smog production (or for that matter air pollution) you might guess that it's the correct answer (and you'd be right!).

We'll explain what these words mean in Chapter 9, in which we discuss pollution, but the point is that without even wracking your brain, you've managed to get this down to two answer choices—not bad! You would have a fifty-fifty chance of guessing correctly on this question.

By combining the associations we'll offer throughout this book with aggressive POE techniques, you'll be able to rack up points on problems that might have seemed particularly difficult at first.

Mnemonics—or the Environmental Science Name Game

One of the big keys to simplifying biology is to organize terms into a handful of easily remembered packages. The best way to accomplish this is by using mnemonics. A mnemonic, as you may already know, is a convenient device, such as a rhyme or phrase, for remembering something. Environmental science is all about names: the names of chemicals, processes, theories, etc. How are you going to keep them all straight without a little help?

For example, the major components of air pollution are

- **s**ulfur dioxide — SO_2

- **p**articulates

- **l**ead — Pb

- **o**zone — O_3

- **n**itrogen dioxide — NO_2

- **c**arbon monoxide — CO

The first letter of each component spells SPLONC, which is otherwise known as Some Pollution Lands On Nature Constantly. Learn the mnemonic and you'll never forget the science!

Mnemonics can be as goofy as you like, so long as they help you remember. Be creative! Remember: The important thing is that you remember the information, not how you remember it.

Identify Question Types

Many of the traps on the AP Environmental Science Exam deal with the way in which the question is asked. For example, one particular type of question you'll see falls under the category of

EXCEPT/NOT/LEAST Questions

> About 10 percent of the multiple-choice questions in Section I are EXCEPT/NOT/LEAST questions. With this type of question, you must remember that you're looking for the wrong (or the least correct) answer. The best way to approach these is by using POE.

More often than not, the correct answer is a true statement, but is wrong in the context of the question. Cross off the four that apply, and you're left with the one that does not. Here's an example of this type of question.

27. All of the following are components of integrated waste management EXCEPT

✗ (A)   Using canvas bags that can be reused rather than disposable bags

✗ (B)   Using old appliances for construction of artificial reefs

(C)   Using disposable diapers instead of cloth diapers

✗ (D)   Using reused glass bottles

✗ (E)   Using planking made from recycled plastic

If you don't remember anything about integrated waste management, you should at least understand that the question is asking about waste. So, which of the choices does *not* deal with a way to reduce or reuse waste? Well, (C) would result in more, and not less, waste; and it is the correct answer. Remember, the best way to answer these types of questions is: Spot all the right statements and cross them off. You'll wind up with the wrong statement, which happens to be the correct answer.

Now, let's turn our attention to the free-response questions.

Chapter 2
How to
Approach
Free-Response
Questions

THE ART OF THE FREE-RESPONSE ESSAY

You're given four essay questions to answer in 90 minutes. That's only 22 minutes per question. Each of these four questions will present a scenario and ask you to answer several smaller questions (generally 3 to 4). You can get a maximum of 10 points per free-response question, and you need to answer every part of the question to get all 10 points. Each question has a certain grading rubric assigned to it, which is what the essay readers use to give you points for your responses. The best way to rack up points on this section is to give the graders what they're looking for. Fortunately, we know precisely how to do this.

Now or Later?

On this test, one of the free-response questions will be a document-based question, or DBQ, and one will be based on data that will be provided to you. The third and fourth questions will be synthesis or analysis questions, and one of those will probably require you to perform some simple calculations.

While you do have to answer each of these questions, you do *not* need to answer them in order. The best strategy is to read the scenario (not the sub-questions) and decide if this is a question you want to attempt now or later. Do this before reading the next scenario. If you decide to do it later, move on to the next question.

If you decide to do it now, look at the questions and start answering them. Remember the grading rubric, and make it easy for the grader to give you points. If the question asks for two solutions, label the solutions (for example, "a" and "b") so the grader can easily find them. If the question asks for a calculation, show all your work, including any formulas you are using. If the question asks you to plot something, clearly label the x and y axes and any relevant point or area on the chart.

Do the Math

Yes, there is math in environmental science. However, you should be able to deal with it easily using a pencil and paper. For example, if you are asked to calculate the cost of heating a 2,000-square-foot house during a Midwestern winter, the answer will be a round number or otherwise easily manipulated figure like $500, not $327.67. Keep in mind that you are not allowed to use a calculator on this test, and remember to provide units with all of your numbers!

Hot-Button Terms

The ETS essay graders have a checklist of key terms and concepts that they use to assign points. We like to call these "hot-button" terms. Quite simply put, for each hot button that you include in your essay, you will receive a predetermined number of points. For example, if the essay question deals with photochemical smog, the ETS graders are instructed to give students two points for writing: "In the presence of sunlight and heat, VOCs (volatile organic compounds), NO_x, and ozone combine to form smog"—or something very similar to that. So where do you find these key terms? Funny you should ask—they are at the end of each chapter in this book, in a section we like to call "Key Terms." Make sure you have a grasp of all of the words in those lists, and use them as hot buttons in your essays.

Make an Outline

As you read the question, brainstorm a list of terms and concepts you want to cover. Use the sub-questions to help you with your list of terms and concepts. Next, draft an outline that will help you organize them into some logical order. While you do not get points for organization, a well-organized essay is easier to write (and more importantly, easier to grade). The best way to organize your response is to write a clear, simple outline (just a couple of bullets per section). Outlining should take no more than about two to three minutes.

Of course, if you just composed a list of key scientific terms, you wouldn't be writing an essay. It is important to remember that the four free-response questions are essay questions, and they need to be written in paragraph style. An answer that's written as a list or an outline is not acceptable and will not be scored. On average, you will need to write no more than one or two paragraphs for each question.

If the question asks for two examples, give just that—two examples. If you present more than two examples, the grader may not even count them toward your score. Make sure you read carefully and provide just what the question asks for.

Label All Diagrams and Figures

Sometimes it's easier to present a diagram or figure as part of your essay. You may illustrate your answer, but all illustrations should be labeled and discussed in the verbiage of your answer. Remember to properly label your diagram or figure; otherwise the ETS graders will give you no more than partial credit for your work.

Know the Labs Covered in Your AP Course

At least one of the four essay questions will be experimentally based. Sometimes the questions will refer back to a laboratory experiment conducted in your AP class. Consequently, the laboratory component of your course is an integral part of this exam. In Chapter 11 we'll review some of the laboratory experiments you may have performed in your AP Environmental Science class.

Review Your Answers

After answering all parts of a question, give yourself a couple of minutes to review your answers before moving to the next question. Remember, once you are done with a question, you are DONE. Do not go back to a question you have completed—even if you have time at the end of the test.

Practice, Practice, and More Practice!

The only way to get good at writing an essay in 22 minutes is to keep at it. Try out the strategies you've learned on the free-response questions at the end of each chapter in Part IV, and on the questions from past exams at AP Central: **https://apstudent.collegeboard.org/apcourse/ap-environmental-science/ exam-practice.**

In the next chapter, we'll give you some additional guidelines to hone your pacing skills.

Chapter 3
Using Time
Effectively
to Maximize
Points

Very few students stop to think about how to improve their test-taking skills. Most assume that if they study hard, they will test well, and if they do not study, they will do poorly. Most students continue to believe this even after experience teaches them otherwise. Have you ever studied really hard for an exam, then blown it on test day? Have you ever aced an exam for which you thought you weren't well prepared? Most students have had one, if not both, of these experiences. The lesson should be clear: Factors other than your level of preparation influence your final test score. This chapter will provide you with some insights that will help you perform better on the AP Environmental Science Exam and on other exams as well.

PACING AND TIMING

A big part of scoring well on an exam is working at a consistent pace. The worst mistake made by inexperienced or unsavvy test takers is that they come to a question that stumps them, and, rather than just skip it, they panic and stall. Time stands still when you're working on a question you cannot answer, and it is not unusual for students to waste five minutes on a single question (especially a question involving a graph or the word EXCEPT) because they are too stubborn to cut their losses. It is important to be aware of how much time you have spent on a given question and on the section you are working on. There are several ways to improve your pacing and timing for the test:

- **Know your average pace.** While you prepare for your test, try to gauge how long you take on 5, 10, or 20 questions. Knowing how long you spend on average per question will help you identify how many questions you can answer effectively and how best to pace yourself for the test.

- **Have a watch or clock nearby.** You are permitted to have a watch or clock nearby to help you keep track of time. It is important to remember, however, that constantly checking the clock is in itself a waste of time and can be distracting. Devise a plan. Try checking the clock after every 15 or 30 questions to see if you are keeping the correct pace or need to speed up. This will ensure that you are cognizant of the time but will not permit you to fall into the trap of dwelling on it.

- **Know when to move on.** Since all questions are scored equally, investing appreciable amounts of time on a single question is inefficient and can potentially deprive you of the chance to answer easier questions later on. If you are able to eliminate answer choices, do so, but don't worry about picking a random answer and moving on if you cannot find the correct answer. Remember, tests are like marathons; you do best when you work through them at a steady pace. You can always come back to a question you don't know. When you do, very often you will find that your previous mental block is gone, and you will wonder why the question perplexed you the first time around (as you gleefully move on to the next question). Even if you still don't know the answer, you will not have wasted valuable time you could have spent on easier questions.

- **Be selective.** You don't have to do any of the questions in a given section in order. If you are stumped by an essay or multiple-choice question, skip it or choose a different one. In the section below, you will see that you may not have to answer every question correctly to achieve your desired score. Select the questions or essays that you can answer and work on them first. This will make you more efficient and give you the greatest chance of getting the most questions correct.

- **Use Process of Elimination on multiple-choice questions.** Many times, one or more answer choices can be eliminated. Every answer choice that can be eliminated increases the odds that you will answer the question correctly. Review the section on this strategy in Chapter 1 to find these incorrect answer choices and increase your odds of getting the question correct.

Remember, when all the questions on a test are of equal value, no one question is that important. Your overall goal for pacing is to get the most questions correct. Finally, you should set a realistic goal for your final score. In the next section, we will break down how to achieve your desired score and ways of pacing yourself to do so.

GETTING THE SCORE YOU WANT

Depending on the score you need, it may be in your best interest *not* to try to work through every question. Check with the schools to which you are applying. Do you need a 3 to earn credit for the test? If you do well on 60 of the multiple-choice questions, plus earn a decent score on at least 3 of the free-response questions, you will get a 3.

Since the administration of the May 2011 exam, AP exams in all subjects no longer include a "guessing penalty" of a quarter of a point for every incorrect answer. Instead, students are assessed only on the total number of correct answers. A lot of AP materials, even those you receive in your AP class, may not include this information. It is really important to remember that if you are running out of time, you should fill in all the bubbles before the time for the multiple-choice section is up. Even if you don't plan to spend a lot of time on every question and even if you have no idea what the correct answer is, it's to your advantage to fill something in.

On the next page is a table to give you an idea of how many questions you need to attempt to get the score you need. Realize that these numbers are approximations and will vary from year to year depending upon test performance. From these data, it becomes readily apparent that you must attempt and perform well on the essays to have a chance to score a 4 or 5. As you take practice tests, you can use this information to evaluate how best to get the score you want and what areas of the exam are hindering your progress. You can calculate your own score on the Practice Tests in this book using the worksheets on pages 322 and 359. There are multiple ways to achieve your desired score. It is important to remember that guessing is no longer penalized and that you must put in the energy and effort on the essays to perform well.

How to Get the Score You Want		
Composite AP Score	**Multiple-Choice Questions**	**Free-Response Questions**
5	All	3
	75	4
4	All	2
	80	3
	60	4
3	All	1
	80	2
	60	3
	50	4

TEST ANXIETY

Everybody experiences anxiety before and during an exam. To a certain extent, test anxiety can be helpful. Some people find that they perform more quickly and efficiently under stress. If you have ever pulled an all-nighter to write a paper and ended up doing good work, you know the feeling.

However, too much stress is definitely a bad thing. Hyperventilating during the test, for example, almost always leads to a lower score. If you find that you stress out during exams, here are a few preemptive actions you can take.

- **Take a reality check.** Evaluate your situation before the test begins. If you have studied hard, remind yourself that you are well prepared. Remember that many others taking the test are not as well prepared, and (in your classes, at least) you are being graded against them, so you have an advantage. If you didn't study, accept the fact that you will probably not ace the test. Make sure you get to every question you know something about. Don't stress out or fixate on how much you don't know. Your job is to score as high as you can by maximizing the benefits of what you do know. In either scenario, it is best to think of a test as if it were a game. How can you get the most points in the time allotted to you? Always answer questions you can answer easily and quickly before you answer those that will take more time.

- **Try to relax.** Slow, deep breathing works for almost everyone. Close your eyes, take a few, slow, deep breaths, and concentrate on nothing but your inhalation and exhalation for a few seconds. This is a basic form of meditation, and it should help you to clear your mind of stress and, as a result, concentrate better on the test. If you have ever taken yoga classes, you probably know some other good relaxation techniques. Use them when you can (obviously, anything that requires leaving your seat and, say, assuming a handstand position won't be allowed by any but the most free-spirited proctors).

- **Eliminate as many surprises as you can.** Make sure you know where the test will be given, when it starts, what type of questions are going to be asked, and how long the test will take. You don't want to be worrying about any of these things on test day or, even worse, after the test has already begun.

The best way to avoid stress is to study both the test material and the test itself. Congratulations! By buying or reading this book, you are taking a major step toward a stress-free AP Environmental Science Exam.

REFLECT

Think about what you've learned in Part III, and respond to the following questions:

- How long will you spend on multiple-choice questions?

- How will you change your approach to multiple-choice questions?

- What is your multiple-choice guessing strategy?

- What will you do before you begin answering a free-response question?

- How will you change your approach to the essays?

- Will you seek further help, outside of this book (such as a teacher, tutor, or AP Central), on how to approach multiple-choice questions, the free-response questions, environmental science labs, or a pacing strategy?

Part IV
Content Review for the AP Environmental Science Exam

HOW TO USE THE CHAPTERS IN THIS PART

You may need to come back to the following chapters more than once. Your goal is to obtain mastery of the content you are missing, and a single read of a chapter may not be sufficient. At the end of each chapter, you will have an opportunity to reflect on whether you truly have mastered the content of that chapter.

Chapter 4
Earth's Interdependent Systems

In this chapter, we will discuss planet Earth's structure and its resources. This review will act as a foundation for many of the chapters that come after it, so pay close attention! According to ETS, about 10–15 percent of the test will be on the material that's contained in this chapter, so consult your textbook for more information about subjects that don't look familiar.

In this chapter, we will begin by discussing the components of the solid Earth—what it's made up of, how long it has existed, why we have earthquakes and volcanoes, and things like that. This will be followed by a discussion of soil—how it is formed, what it's composed of, and why it's important to humans. We'll move on to discuss the atmosphere—its structure and how it produces the weather we experience. The atmosphere, hydrosphere, and biosphere are all intricately related, and can almost not be discussed separately. However, we've broken them down into neat sections for review, and we'll go through everything you'll need to know about these systems for test day.

We'll end this chapter by reviewing oceans and freshwater bodies; we'll refresh your memory about certain environmental issues that pertain to Earth's water supply, too.

WELCOME TO PLANET EARTH

The first thing you should know about Earth is its history. Earth is thought to be between 4.5 and 4.8 billion years old. That amount of time is pretty inconceivable to humans, but the geologic time scale on the following page will help you get a sense of the vast amount of time that has gone by since Earth was formed. You will not be responsible for memorizing all of the eons, eras, periods, and epochs for this exam, but you should be familiar with the major ones; they will come in handy.

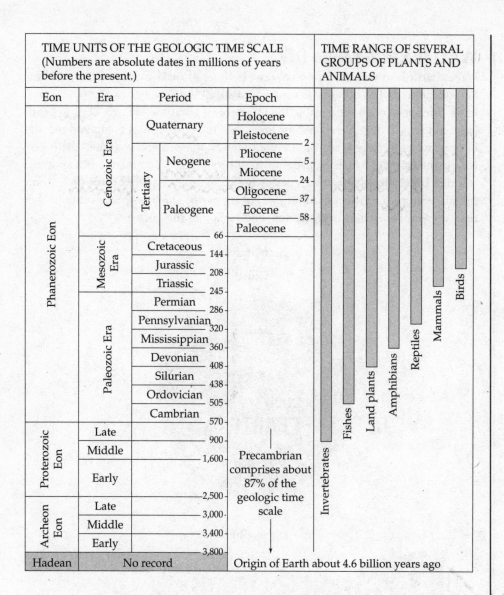

TIME UNITS OF THE GEOLOGIC TIME SCALE (Numbers are absolute dates in millions of years before the present.)				TIME RANGE OF SEVERAL GROUPS OF PLANTS AND ANIMALS
Eon	Era	Period	Epoch	
Phanerozoic Eon	Cenozoic Era	Quaternary	Holocene	
			Pleistocene — 2	
		Tertiary — Neogene	Pliocene — 5	
			Miocene — 24	
		Tertiary — Paleogene	Oligocene — 37	
			Eocene — 58	
			Paleocene	
			— 66	
	Mesozoic Era	Cretaceous — 144		
		Jurassic — 208		
		Triassic — 245		
	Paleozoic Era	Permian — 286		
		Pennsylvanian — 320		
		Mississippian — 360		
		Devonian — 408		
		Silurian — 438		
		Ordovician — 505		
		Cambrian — 570		
Proterozoic Eon	Late	— 900		
	Middle — 1,600			Precambrian comprises about 87% of the geologic time scale
	Early — 2,500			
Archeon Eon	Late — 3,000			
	Middle — 3,400			
	Early — 3,800			
Hadean	No record			Origin of Earth about 4.6 billion years ago

Time range groups of plants and animals: Invertebrates, Fishes, Land plants, Amphibians, Reptiles, Mammals, Birds

Where Is Earth in the Solar System?

Earth is the third planet from the sun in our solar system, which contains a total of eight planets. From the sun outward, the planets are: Mercury, Venus, Earth, Mars, Jupiter, Saturn, Uranus, and Neptune.

You probably already know that Earth and the other planets orbit the sun in an elliptical pattern. And, of course, it takes Earth about $365\frac{1}{4}$ days, or 1 year, to complete its orbit of the sun.

Memorize the order and characteristics of Earth's layers; they will be asked about on test day!

What Is Earth Made Of?

Planet Earth is made up of three concentric zones of rocks that are either solid or liquid (molten). The innermost zone is the **core**. The core has two parts: a solid inner core and molten outer core. The inner core is composed mostly of nickel and iron, and is solid due to tremendous pressures. The outer core is composed mostly of iron and sulfur, and is semi-solid due to lower pressures. Surrounding the outer core is the **mantle**, which is made mostly of solid rock. The mantle has an area, called the **asthenosphere**, which is slowly flowing rock. The **lithosphere**, a thin, rigid layer of rock, is the outermost layer of the earth. The lithosphere contains the rigid upper mantle and the **crust**, our solid surface of the earth.

Earth's Layers

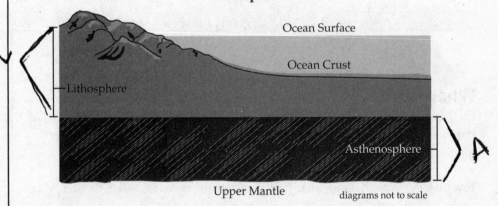

diagrams not to scale

Tectonic Plates

Because the lithosphere floats atop the asthenosphere like a cracker on a layer of hot pudding, it can move and break into large pieces or **tectonic plates.** There are a total of a dozen or so tectonic plates in the lithosphere that move independently of one another. The plates are made up of both mantle and crust. The majority of the land on Earth sits above six giant plates; the remainder of the plates lie under the ocean as well as the continents.

Some plates consist only of ocean floor, such as the Nazca plate, which lies off the west coast of South America, while others contain both continental and oceanic material. One example of the latter is the North American plate, where the United States is located; this plate extends out to the mid-Atlantic ridge. There is even a plate that is located exclusively within the Asian continent; its boundaries nearly coincide with those of Turkey. The largest plate is the Pacific plate—it primarily consists of ocean floor, but also includes Mexico's Baja Peninsula and southwestern California. The major plates of Earth are shown on the map below.

Earth's Plates

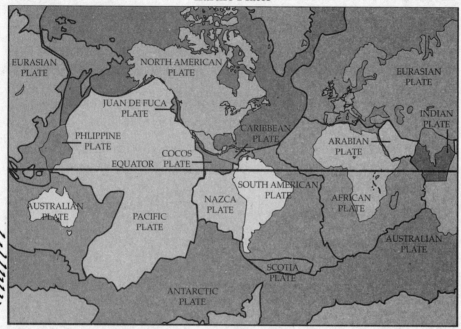

The edges of the plates are called **plate boundaries** and the places where two plates abut each other is where events like sea floor spreading and most volcanoes and earthquakes occur. There are three types of plate boundary interactions.

- **Convergent boundary**: Two plates are pushed toward each other. One of the plates will be pushed deep into the mantle.

- **Divergent boundary**: Two plates are moving away from each other. This causes a gap that can be filled with magma (molten rock), and when it cools new crust is formed.

- **Transform fault boundary:** Two plates slide from side to side relative to each other—like when you rub your hands back and forth. These are also called transform boundaries.

So, what happens when plates collide? It depends on what types of plates collide, and where. Converging ocean-ocean and converging ocean-continent boundaries often result in **subduction,** in which a heavy ocean plate is pushed below the other plate and melts as it encounters the hot mantle. Converging continent-continent boundaries result in the uplifting of plates to form large mountain chains, like the Himalayas (which were created by a collision between the plate carrying India and the Asian plate), the Urals, the Alps, and the Appalachian Mountains.

*Plates collide → creates mountains, volcanoes, or earthquakes.

One important result of plate movement is the creation of volcanoes and earthquakes. Let's examine those next.

Volcanoes and Earthquakes

Volcanoes are mountains formed by magma from the earth's interior. **Active volcanoes** are those that are currently erupting or have erupted within recorded history, while **dormant volcanoes** have not been known to erupt. It's thought that **extinct volcanoes** will never erupt again.

Volcanoes form where tectonic plates meet. At these junctures, breaks occur in the earth's crust and magma flows out. If no outlet is available as the plates push together, pressure builds up until it is relieved in an explosion—a volcanic eruption.

Active volcanoes are categorized by the kind of tectonic event that produces them. The three types are as follows:

- **Rift volcanoes** occur when plates move away from each other. When a rift volcano erupts, new ocean floor is formed as magma fills in where the plates have separated.

- **Subduction volcanoes** occur where plates collide and slide over each other.

- **Hot spot volcanoes** do not form at the margin of plates. Instead, they are found over "hot spots," which are areas where magma can rise to the surface through the plates. The Hawaiian Islands are thought to have formed over a hot spot.

Earthquakes are the result of vibrations (often due to sudden plate movements) deep in the earth that release energy. Earth's tectonic plates move slowly all the time, about the same pace as fingernail growth, but earthquakes are very sudden movements. They often occur as two plates slide past one another at a transform boundary. The **focus** of the earthquake is the location at which it begins within the earth, and the initial surface location is the **epicenter.** The size, or magnitude, of earthquakes is measured by using an instrument known as a **seismograph,** which was devised by Charles Richter in 1935. The Richter scale measures the amplitude of the highest S-wave of an earthquake. Each increase in Richter number corresponds to an increase of approximately 33 times the energy of the previous number.

In January 2010, an earthquake of magnitude 7.0 struck the nation of Haiti. The center of the quake occurred in the boundary region separating the Caribbean plate and the North American plate. Official estimates from the U.S. Geological Service had 222,570 people killed, 300,000 injured, 1.3 million displaced, 97,294 houses destroyed and 188,383 damaged across the Port-au-Prince area and much of southern Haiti. This includes at least 4 people killed by a local tsunami in the Petit Paradis area near Léogâne. Tsunamis are very large waves, or chains of waves, caused by the movement of the earth during an earthquake or volcanic eruption and can be extremely destructive.

In March 2011, Japan suffered an earthquake of 9.0 magnitude off the eastern coast, near Sendai. This quake caused a massive tsunami wave that was 33 feet high. Both the earthquake and tsunami caused extensive damage: Many buildings, roads, and railways were destroyed; major fires occured; villages, thousands of homes, and people were washed away; and at least three nuclear power plants experienced dangerous explosions.

The Rock Cycle

Rocks are all around us, in the soil, our buildings, and the ore used in industry. So, where do all those rocks come from? The answer is: other rocks. The oldest rocks on Earth are 3.8 billion years old, while others are only a few million years old. This means that rocks have to be recycled. The process that does this is the **rock cycle**. In the rock cycle, time, pressure, and the earth's heat interact to create three basic types of rocks.

- **Igneous**—this type of rock results when rock is melted (by heat and pressure below the crust) into a liquid and then resolidifies. The molten rock (**magma**) comes to the surface of the earth, and when it emerges it is called lava; solid lava is igneous rock. An example of an igneous rock is basalt.

- **Sedimentary**—these rocks are formed as sediment (eroded rocks and the remains of plants and animals) builds up and is compressed. One place this can occur is at a subduction zone where ocean sediments are pushed deep into the earth and compressed by the weight of rock above it. An example of a sedimentary rock is limestone.

- **Metamorphic**—this type of rock is formed as a great deal of pressure and heat is applied to rock. This can happen as sedimentary rocks sink deeper into the earth and are heated by the high temperatures found in the earth's mantle. An example of a metamorphic rock is slate (it is product of metamorphosis of shale).

The diagram on the next page illustrates the rock cycle. Make sure you are familiar with it before the exam!

The Rock Cycle

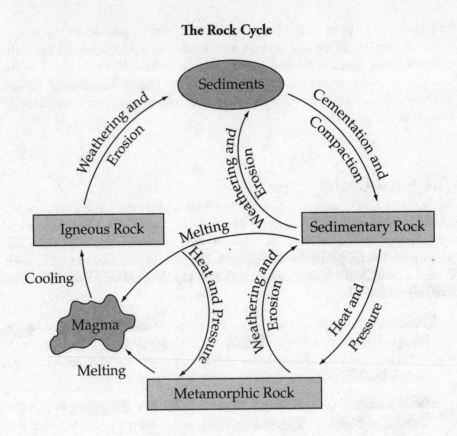

SOIL

One very important but often underappreciated player in Earth's interdependent systems is soil. Soil plays a huge and crucial role in the lives of the plants, animals, and other organisms that live in the biosphere, and acts as a crucial link between the **abiotic** (the nonliving components of the world) and the **biotic** (that's right, the living components of the world). As we'll see in the next chapter, soil also plays an active role in the cycling of nutrients. Let's take a moment to review the major characteristics of soil that you'll be expected to know for the test.

Soil Is More than Just Dirt

Although we may be tempted to think of soil as simply "dirt," soil is actually a complex, ancient material teeming with living organisms. Some soil is hundreds of years old! In just one gram of soil, there may be as many 50,000 protozoa, as well as bacteria, algae, fungi, and larger organisms such as earthworms and nematodes. About one half of the volume of soil is made up of mineral materials, and about 5 percent is organic matter (both living and dead). The pores between the grains of minerals in soil are filled with air or water and, as a rule, the size of the particles that make up the soil determines the size of the pores between the soil particles.

Soils can be categorized based on numerous physical and chemical features including color and texture. The United States Department of Agriculture (USDA) divides soil textures into three large groups: The category with the smallest particles is **clay,** which has particles that are less than 0.002 mm in diameter. The next largest is **silt,** with particles 0.002–0.05 mm in diameter, and **sand** is the coarsest soil, with particles 0.05–2.0 mm in diameter. Sand particles are too large to easily stick together, and sandy soils have larger pores; which means that they can hold more water. Clays easily adhere to each other and there is little room between particles for water; clay soil is extremely compact.

Another very important characteristic of soil types is soil **acidity** or **alkalinity.** Recall that the pH of a substance ranges from 0–14, and is a measure of the concentration of hydrogen ions. Most soils fall into a pH range of about 4–8, meaning that most soils range in pH from being neutral to slightly acidic. Soil pH is important because it affects the solubility of nutrients; and this in turn determines the extent to which these nutrients are available for absorption by plant roots. If the soil in a region is too acidic or basic, certain soil nutrients will not be able to be used by the regional plants. One last thing about soils that the College Board wants you to know: when the pH of the soil gets more acidic, ions of heavy metals such as mercury (Hg) or aluminum (Al) can leach into the ground water. These ions can travel to streams and rivers and harm both plants and aquatic life. For example, aluminum ions can damage the gills of fish and cause them to suffocate.

Where Does Soil Come From?

Basically, soil is a combination of organic material and rock that has been broken down by chemical and biological weathering. Therefore, it should not be surprising to learn that the types of minerals found in soil in a particular region will depend on the identity of the base rock of that region.

Water, wind, and living organisms are all prominent agents of weathering, and all weathering processes are placed into the following three rather broad categories (which you should definitely know for exam day).

- **Physical weathering** (also known as **mechanical weathering**): Any process that breaks rock down into smaller pieces without changing the chemistry of the rock. The forces responsible for physical weathering are typically wind and water.

- **Chemical weathering:** Occurs as a result of chemical interactions between water and other atmospheric gases, and the bedrock of a region. One example of chemical weathering is rust, which forms when iron and other metallic elements come in contact with water.

- **Biological weathering:** Weathering that takes place as the result of the activities of living organisms, such as tree roots growing and expanding through rocks.

Soil is made up of distinct layers with very different characteristics. Let's discuss those next.

Soil Layers

Soil comprises distinct layers known as **horizons,** which vary considerably in content.

- **O horizon:** The O horizon is the uppermost horizon of soil. It is primarily made up of organic material, including waste from organisms; the bodies of decomposing organisms; and live organisms. The dark, crumbly material that results from the decomposition of organic material forms **humus.**

- **A horizon:** The horizon below the O layer is called the A horizon, and this layer is made up of weathered rock and some organic material that has traveled down from the O layer. The A layer is often referred to as **topsoil** and plays an important role in plant growth. This is the zone of **leaching**.

- **E horizon:** In between the A & B horizons, the E horizon, or eluviated horizon, is one that has been leached of clay, iron, or aluminum oxides; minerals in this layer are sand and silt sized.

- **B horizon:** This layer lies below the A horizon. The B layer receives all of the minerals that are leached out of the A horizon as well as organic materials that are washed down from the topsoil above. This is the zone of **illuviation**. Illuviation is the movement of dissolved material from higher soil layers to lower soil layers due to the downward movement of water, which is caused by gravity.

- **C horizon:** The bottommost layer of soil is the C horizon. The C horizon is composed of larger pieces of rock that have not undergone much weathering.

- **R horizon:** The bedrock, which lies below all of the other layers of soil, is referred to as the R horizon.

Memorize the different types of weathering processes and the order and characteristics of soil layers. You will definitely be asked about these on test day!

Soil Problems for (and Caused by) Humans

In order to be able to grow all of the foods that humans consume, we must have enough **arable**—or suitable for plant growth—soil to meet our agricultural needs. Soil fertility refers to soil's ability to provide essential nutrients, like nitrogen (N), potassium (K), and phosphorus (P), to plants. Humus (remember, it's in the O layer!) is also an extremely important component of soil because it is rich in organic matter.

Soils composed of roughly the same amount of all three textures (remember: clay, silt, and sand) are described as being **loamy,** and these types of soil are considered the best for plant growth. Another important characteristic of soil for agricultural purposes is the extent to which it aggregates, or clumps. The most fertile soils are aggregates (look, it's a noun, too!) of soils of different textures bound together with organic material.

Monoculture

Unfortunately, certain agricultural activities can change the texture of soil; for example, repeated plowing tends to break down soil aggregates, leaving "plow pan" or "hard pan," which is hard, unfertile soil.

Whereas communities traditionally planted many different types of crops in a field, in modern agriculture the **monoculture,** or the planting of just one type of crop in a large area, predominates. Over the history of agriculture, a significant decrease in the genetic diversity of crop species has taken place. This creates numerous problems. First of all, a lack of genetic variation makes crops more susceptible to pests and diseases. Secondly, the consistent planting of one crop in an area eventually leaches the soil in that area of the specific nutrients that the plant needs in order to grow. One way to prevent this phenomenon is to practice **crop rotation,** in which different crops are planted in the area in each growing season.

Other problems with modern agriculture include its reliance on large machinery (which can damage soil), and the fact that as an industry, agriculture is a huge consumer of energy. Energy is consumed both in the production of pesticides and fertilizers and in the use of fossil fuels to run farm machinery.

The past 50 years or so have seen a huge increase in worldwide agricultural productivity, which is largely due to the mechanization of farming that resulted from the Industrial Revolution. The boom in agricultural productivity is known as the **Green Revolution,** and, unfortunately, it has since had many detrimental effects on the environment. For example, the use of chemical pesticides resulted in the emergence of new variety of insects that were pesticide-resistant. Recently, the introduction of genetically modified plants has enabled researchers to take steps in solving the problem of pesticide-resistant insect species.

Another drawback to the Green Revolution resulted from the dramatic increase in irrigation worldwide; over-irrigated soils undergo salinization. In **salinization,** the soil becomes water-logged and when it dries out, salt forms a layer on its surface; this eventually leads to **land degradation.** In order to combat this problem, researchers have developed **drip irrigation,** which allots an area only as much water as is necessary and delivers the water directly to the roots.

Soil Erosion

As you learned earlier, the small rock fragments that result from weathering may be moved to new locations in the process of erosion, and bare soil (soil upon which no plants are growing) is more susceptible to erosion than soil that's covered by organic materials.

Because of the constant movement of water and wind on Earth's surface, the **erosion** of soil is a continual and normal process. However, when erosion removes valuable topsoil or deposits soil in undesirable places, it can become a problem for humans. Eroded topsoil usually ends up in bodies of water, posing a problem for both farmers, who need healthy soil for planting, and people who rely on bodies of water to be uncontaminated with soil runoff.

The most significant portion of erosion caused by humans results from logging and slash-and-burn agriculture, which is characterized by cutting down and burning trees to clear land for agricultural purposes. The removal of plants in an area makes the soil much more susceptible to the agents of erosion.

Unfortunately, human activities such as the ones we have just discussed—the over-cultivation of agricultural fields, overgrazing, urbanization, and deforestation—have significantly increased the levels of erosion in the upper layers of soil. These processes will continue to create problems for farmers searching for arable land until new techniques that preserve the integrity of soil are introduced and utilized.

Soil Conservation

In order to conserve soil resources, several best management practices have been developed. These practices return organic matter to the soil, slow down the effects of wind, and reduce the amount of damage done to the soil by tillage (plowing). Here are some of the more common methods.

- Use of animal waste (manure), compost, and the residue of plants to increase the amount of organic matter in the soil.

- Practice of organic agriculture, a method of farming that utilizes compost, manure, crop rotation, and non-chemical methods to manage soil fertility and pest control. Organic producers do not use, or strictly limit, the use of chemical fertilizers and pesticides or genetically modified organisms.

- Modification of tillage practices to reduce the breakup of soil and to reduce the amount of erosion. These include contour plowing and strip planting.

- Use of trees and other wind barriers to reduce the force of the wind.

Much of what we know about soil conservation was established relatively recently. The Soil Conservation Act was passed in 1935 and lead to the creation of the Soil Conservation Service. These were created in response to the Dust Bowl of the 1930s, which was a period of unprecedented dust storms caused by severe drought and ill-advised farming practices. This program was created to conserve soil and restore the nation's ecological balance and was lead by soil conservation pioneer Hugh Hammond Bennett.

Soil Laws

The federal government recognized the need to protect this vital resource. Review the two laws below that relate to preserving soil.

Date	Name of Law	What It Does
1977	Soil and Water Conservation Act	Soil and water conservation programs to aid landowners and users; also sets up conditions to continue evaluating the condition of U.S. soil, water, and related resources.
1985	Food Security Act	Nicknamed the Swampbuster, this act discouraged the conversion of wetlands to nonwetlands. 1990 federal legislation denied federal farm supplements to those who converted wetlands to agriculture, and provided a restoration of benefits to those who converted lands to wetlands.

Phew, got it? You're almost at the halfway point of this chapter. If you want, go get a glass of water or a snack to bolster you for the remaining pages!

Next we will turn to a study of what surrounds the earth: the atmosphere.

THE ATMOSPHERE

In the broadest definition, the atmosphere is a layer of gases that's held close to the earth by the force of gravity. The layer of gases that lies closest to the earth is the **troposphere;** it extends from the surface of the earth to about 10–20 km (5–10 miles). The troposphere is where all of the weather that we experience takes place; this layer contains the majority of atmospheric water vapor and clouds. Generally the troposphere is vertically well mixed and (with the exception of periods of temperature inversions) it gradually becomes colder with an increase in altitude (by about 6.5°C/km).

Make sure you have a solid grasp of the order and characteristics of the layers of Earth's atmosphere!

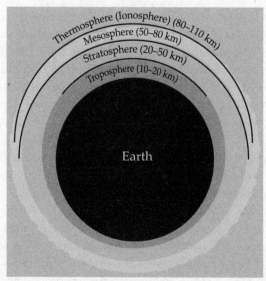

Thermosphere (Ionosphere) (80–110 km)
Mesosphere (50–80 km)
Stratosphere (20–50 km)
Troposphere (10–20 km)

Earth

diagram not to scale

You've probably heard about the troposphere before in the news because of the **greenhouse effect.** The troposphere contains certain gases called "greenhouse" gases, the most important of which are H_2O and CO_2. As the sun's rays strike the earth, some of the solar radiation is reflected back into space; however, greenhouse gases in the troposphere intercept and absorb a lot of this radiation. We'll further investigate the greenhouse effect in Chapter 9, but for now take a look at the following figure.

The Greenhouse Effect

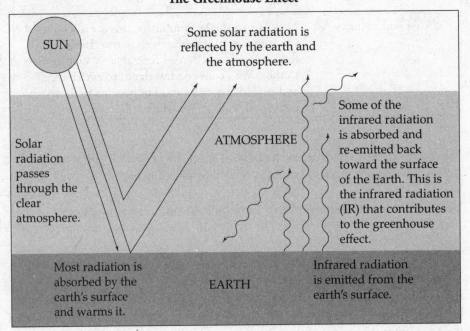

Crowning the troposphere is the **tropopause,** which is a layer that acts as a buffer between the troposphere and next layer up, the stratosphere. In this buffer zone, the atmospheric temperature no longer decreases with altitude; instead, the temperature begins to *increase* with altitude.

The **stratosphere** sits on top of the tropopause and extends about 20–50 km above the earth's surface. Unlike the troposphere, its gases are not very well-mixed and, as in the tropopause, the temperature in the stratosphere increases as the distance from Earth increases. This warming effect is due to a thin band of **ozone** (O_3) that exists in this layer. The ozone traps the high-energy radiation of the sun, holding some of the heat and protecting the troposphere and the earth's surface from this radiation.

Above the stratosphere are two layers called the mesosphere and the thermosphere (ionosphere). The **mesosphere** extends about 80 km above the earth's surface and is the area where meteors usually burn up.

The **thermosphere** is the thinnest gas layer; it is located about 110 km above Earth and is where auroras take place. It's also the layer of the atmosphere where space shuttles orbit! The thermosphere is also known as the **ionosphere** because of the ionization that takes place in this region; this region also absorbs most of the energetic charged particles, such as protons and electrons (solar wind), from the sun. Interestingly, the thermosphere also reflects radio waves, which is what makes long distance radio communication possible. You'll need to know how the climates that we experience on Earth are created by the atmosphere, so let's go into this next.

Climate

The earth's atmosphere has physical features that change day to day as well as patterns that are consistent over a space of many years. The day-to-day properties such as wind speed and direction, temperature, amount of sunlight, pressure, and humidity are referred to as **weather**. The patterns that are constant over many years (30 years or more) are referred to as **climate**. The two most important factors in describing climate are average temperature and average precipitation amounts. **Meteorologists** are scientists who study weather and climate.

The weather and climate of any given area is the result of the sun unequally warming the earth (and the gases above it) as well as the rotation of the earth.

Air Circulation in the Atmosphere

The motion of air around the globe is the result of solar heating, the rotation of Earth, and the physical properties of air, water, and land. There are three major reasons that Earth is unevenly heated.

- More of the sun's rays strike the earth at the equator in each unit of surface area than strike the poles in the same unit area.

- The tilt of Earth's axis points regions toward or away from the sun. When pointed toward the sun those areas receive more direct or intense light than when pointed away. This causes the seasons.

- Earth's surface at the equator is moving faster than the poles. This changes the motion of air into major **prevailing winds**, belts of air that distribute heat and moisture unevenly. Winds moving north from the equator in the Northern Hemisphere are deflected to the right or east, and winds moving south from the equator in the Southern Hemisphere are deflected to the left or east. This deflection pattern is know as the Coriolis effect.

Solar energy warms Earth's surface. The heat is transferred to the atmosphere by radiation heating. The warmed gases expand, become less dense, and rise creating vertical currents called **convection currents**. The warm currents can also hold a lot of moisture compared to the surrounding air. As these large masses of warm moist air rise, cool air flows along Earth's surface into the area where the warm air was located. This flowing air or **horizontal airflow** is one way that surface winds are created.

As warm moist air rises into the cooler atmosphere, it cools to the **dew point**, the temperature at which water vapor condenses into liquid water. This condensation creates clouds. If condensation continues and the drops get bigger, they can no longer be held up by the convection in the earth's atmosphere and they fall as **precipitation** (which can be frozen or liquid). The cold dry air is now denser than the surrounding air. This air mass then sinks to the earth's surface where it is warmed and can gather more moisture, thus starting the **convection cell** rotation again.

Convection Cell

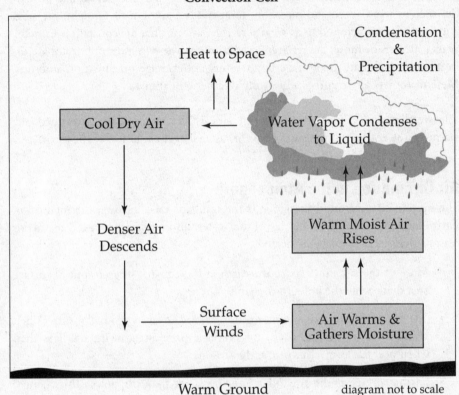

On a local level this phenomena accounts for land and sea breezes. On a global scale these cells are called **Hadley cells**. A large Hadley cell starts its cycle over the equator, where the warm moist air evaporates and rises into the atmosphere. The precipitation in that region is one cause of the abundant equatorial rainforests. The cool dry air then descends about 30 degrees north and south of the equator, forming the belts of deserts seen around the earth at those latitudes.

Hadley Cell

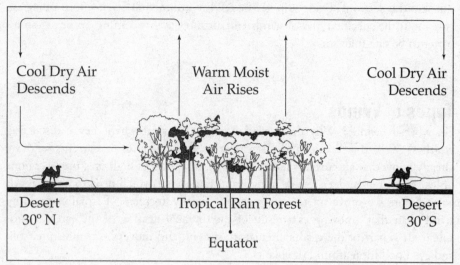

diagram not to scale

Seasons

The motion of Earth around the Sun—and the fact that Earth is tilted on its axis by 23.5 degrees—together create the seasons that we experience on Earth. When Earth is in the part of its orbit in which the Northern Hemisphere is tilted toward the sun, the northern half of the planet receives more direct sunlight for longer periods of time each day than does the Southern Hemisphere. This means that when the Northern Hemisphere is experiencing summer, the Southern Hemisphere is experiencing winter.

Interestingly, because of the earth's tilt, the sun rises and sets just once a year at the North and South Poles. Approximately six months of the year at the poles are daytime, while the other six months are dark, and considered nighttime.

Seasons

diagram not to scale

Let's move on to discuss wind. You might be thinking, "I know what wind is, so I can skip this section!" Don't skip it! The AP Environmental Science Exam may ask you about the specific types of winds and air movements coming up next—so it's better to be safe than sorry.

Types of Winds

So, what is "wind"? Why does everyone refer to wind when they're discussing weather? Well, the term "wind" is widely used to refer to air currents, and we already know that air currents tend to flow from regions of high pressure to regions of low pressure. But let's review some important details you'll need to know about wind before we move on to our review of the hydrosphere. Formally speaking, **wind** is air that's moving as a result of the unequal heating of the earth's atmosphere. It is part of the earth's circulatory system, and moves heat, moisture, soil, and even pollution around the planet.

One crucial wind-related phenomenon that you'll need to know about for the AP Environmental Science Exam is trade winds. **Trade winds** were named for their ability to quickly propel trading ships across the ocean. The trade winds that blow between about 30 degrees latitude and the equator are steady and strong, and travel at a speed of about 11 to 13 mph. In the Northern Hemisphere, the trade winds blow from the northeast and are known as the **Northeast Trade Winds;** in the Southern Hemisphere, the winds blow from the southeast and are called the **Southeast Trade Winds.**

Another important type of moving air mass, called a **westerly,** travels south and west in the Northern Hemisphere and north and west in the Southern Hemisphere near the equator (between 30 degrees and 60 degrees). The movement of air that accounts for the westerlies, called the Ferrel cell, is the reverse of the Hadley cell. Westerlies are another result of the Coriolis effect. **Polar easterlies** are formed by similar forces; in polar easterlies, winds between latitudes 60 degrees and the North Pole blow from the north and east, and winds between 60 degrees and the South Pole blow from the south and east.

Between the wind belts mentioned above, air movement is less predictable, and often no wind blows at all for days. For example, between about 30 degrees to 35 degrees north and 30 degrees to 35 degrees south of the equator lies the region known as the **horse latitudes** (or the subtropical high). This region of subsiding dry air and high pressure results in very weak winds. Some people say that sailors gave the region of the subtropical high the name "horse latitudes" because ships relying on wind were unable to sail in these areas—afraid of running out of food and water, sailors threw their horses (and other live cargo) overboard to save on food and water and to make the ship lighter and easier to move.

Similarly, the air near the equator is relatively still because air at these locations is constantly rising, and not blowing. For this reason, early sailors called this region the **doldrums.** The doldrums, which exist between 5 degrees north and 5 degrees south of the equator, are also known as the Intertropical Convergence

Zone, or ITCZ for short. The trade winds converge in the region of the ITCZ, producing convectional storms that produce regions with some of the world's heaviest precipitation.

The last type of moving air system that you'll need to be familiar with for the exam is the **jet stream.** Jet streams are high-speed currents of wind that occur in the upper troposphere; these fast-moving air currents have a large influence on local weather patterns.

It's important that you can distinguish among the different types of wind and blow through the wind questions on test day!

Winds Around The World

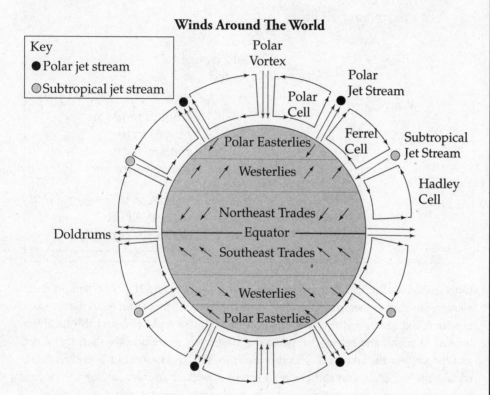

Got it? Let's move on to review the types of weather that result from all of these moving air masses.

Weather Events

Let's start by discussing the monsoon. **Monsoons**, or seasonal winds that are usually accompanied by very heavy rainfall, are caused by the fact that land heats up and cools down more quickly than water does. In a monsoon, hot air rises from the heated land, and a low-pressure system is created. The rising air is quickly replaced by cooler moist air that blows in from over the ocean. As this air rises, it cools, and the moisture it carries is released in a steady seasonal rainfall. This process happens in reverse in the dry season, when masses of air that have cooled over the land blow out over the ocean. Monsoons primarily occur in coastal areas. Check out this helpful illustration.

How a Monsoon Forms

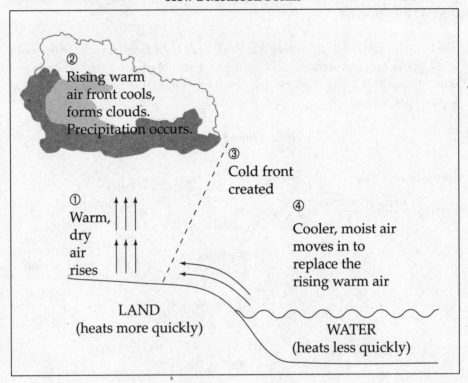

On a smaller scale, this effect can be seen on the shores of large lakes or bays. In these areas, again the land warms faster than does the water during the day, so the air mass over the land rises. Air from over the lake moves in to replace it, and this creates a breeze. At night, the reverse happens; the land cools more quickly than the water and the air over the lake rises. The air mass from the land moves out over the lake to replace the rising air, and this creates a breeze as well. If you live in Chicago or San Francisco, you may have experienced this small-scale monsoon effect!

As we mentioned above, the air that moves in from over the ocean or a large body of water contains large amounts of water. If an air mass is forced to climb in altitude—if, for instance, it encounters an obstruction such as a mountain, the air will be forced to rise. When the air mass rises, it will cool and water will precipitate out on the ocean side of the mountain. By the time the air mass reaches the opposite side of the mountain, it will be virtually devoid of moisture. This phenomenon is known as the **rain shadow effect,** and is responsible for the impressive growth of the Olympic rainforest on the Washington State coast. Interestingly, the Olympic rainforest receives up to 5 m of rain per year, while the opposite side (the leeward side) receives less than 50 cm of rain per year.

How The Rainshadow Effect Works

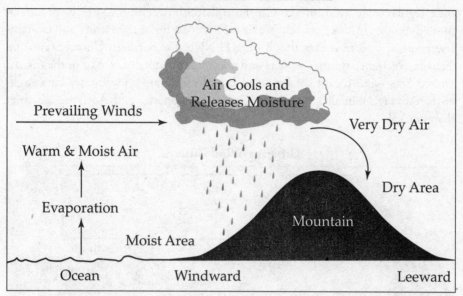

Remember trade winds from the last section? Well, they occur in steady and somewhat predictable wind patterns, but they may cause local disturbances when they blow over very warm ocean water. When this occurs, the air warms and forms an intense, isolated, low-pressure system, while also picking up more water vapor from the ocean surface. The wind will circle around this isolated low-pressure air area (counterclockwise in the Northern Hemisphere and opposite in the Southern Hemisphere—once again because of the Coriolis effect!). The low pressure system will continue to move over warm water, increasing in strength and wind speed; this will eventually result in a tropical storm.

Certain tropical storms are of sufficient intensity to be classified as **hurricanes.** Hurricanes can have winds with speeds in excess of 130 km/hr. The rotating winds of a hurricane remove water vapor from the ocean's surface, and heat is released as the water vapor condenses. This addition of heat energy continues to contribute to the increase in wind speed, and some hurricanes have winds traveling at speeds of nearly 400 km per hour! A major hurricane contains more energy than that released during a nuclear explosion, but since the force is released more slowly, the damage is generally less concentrated. Another important note about this type of storm is that they are referred to as hurricanes in the Atlantic Ocean, but they are called **typhoons** or cyclones when they occur in the Pacific Ocean. Go figure!

El Niño is a climate variation that takes place in the tropical Pacific once about every three to seven years, and it lasts for about one year. Under normal weather conditions, trade winds move the warm surface waters of the Pacific away from the west coast of Central and South America. As a result, the cold ocean water that lies under the displaced water moves to the surface (causing the thermocline to rise), bringing nutrients with it and keeping the temperature of the coastal water relatively cool. This phenomenon is called upwelling.

During El Niño, the normal trade winds are weakened or reversed because of a reversal of the high and low pressure regions on either side of the tropical Pacific.

This reversal of pressure systems is known as the **Southern Oscillation.** Without these regular trade winds off the Central and South American coast, the process of upwelling slows or stops, and the water off the coast becomes warmer and contains fewer nutrients. This means that during El Niño, the northern United States and Canada experience warmer winters and a less intense hurricane season; the eastern United States and regions of Peru and Ecuador that are typically dry have higher-than-average rainfall; and the Philippines, Indonesia, and Australia are drier than normal.

Differing Wind Patterns

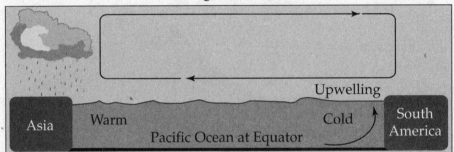

Normal Conditions

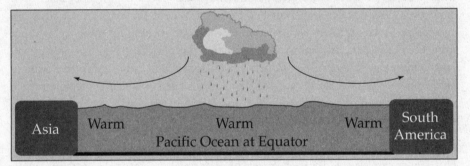

El Niño Conditions

One environmentally important effect that El Niño has on humans is that, because of the suppression of upwelling, the offshore fish populations of certain coastal areas decline. In countries like Peru, which relies heavily on fishing, El Niño has devastating economic effects.

The reverse of El Niño is known as **La Niña.** The Coriolis effect contributes to La Nina conditions. As air moves toward the equator to replace rising hot air, the moving air deflects to the west and helps move the surface water, allowing the upwelling. During La Niña, the surface waters of the ocean surrounding Central and South America are colder than normal. Finally, the alternations of atmospheric conditions that lead to El Niños or La Niñas are referred to as **ENSO** events. (This stands for El Niño and Southern Oscillation.).

THE HYDROSPHERE

Water covers about 75 percent of planet Earth. Most of the water on Earth's surface is salt water. On average, the salt water in the world's oceans has a salinity of about 3.5 percent. This means that for every 1 liter (1,000 ml) of sea water, there are 35 grams of salts (mostly, but not entirely, sodium chloride) dissolved in it. In fact, one cubic foot of seawater would evaporate to leave about 2 pounds of sea salt! However, **sea water** is not uniformly saline throughout the world. The planet's freshest sea water is in the Gulf of Finland, part of the Baltic Sea. The most saline open sea is the Red Sea, where high temperatures and confined circulation result in high rates of surface evaporation.

Freshwater is water that contains only minimal quantities of dissolved salts, especially sodium chloride. All freshwater ultimately comes from precipitation of atmospheric water vapor, which reaches inland lakes, rivers, and groundwater bodies directly, or after melting of snow or ice. Let's start with a discussion of freshwater before discussing the world's oceans. We will end this section with a review of the ways in which humans use water, global problems associated with water usage, and issues of water conservation.

Freshwater

Freshwater is deposited on the surface of the earth through precipitation. Water that falls on the earth and doesn't move through the soil to become groundwater moves along the earth's surface, via gravity, forms small streams, and then eventually forms larger ones. The size of the stream will continue to increase as water is added to it, until the stream becomes a river, and the river will flow until it reaches the ocean. The land area that drains into a particular stream is known as a **watershed,** or drainage basin.

As water moves into streams, it carries with it sediment and other dissolved substances, including small amounts of oxygen. Turbulent waters are especially laden with dissolved oxygen and carbon dioxide, such as those found at the source, or head waters, of a stream. As a general rule, the more turbulent the water, the more dissolved gases it will contain.

Freshwater Bodies

As you probably know, freshwater that travels on land is largely responsible for shaping the earth's surface. Erosion occurs when the movement of water etches channels into rocks. The moving water then carries eroded material farther downstream.

Because of obstructions on land, moving water does not move in a straight line; instead it follows the lowest topographical path, and as it flows it cuts farther into its banks to eventually form a curving channel. As the water travels around these bends, its velocity decreases and the stream drops some of its sedimentary load. Water always follows the path of least resistance as it travels from the highlands to the sea.

Rivers drop most of their sedimentary load as they meet the ocean because their velocity decreases significantly at this juncture. At these locations, landforms called **deltas** (which are made of deposited sediments) are created. Another important freshwater body that you should know about is the estuary. **Estuaries** are sites where the "arm" of the sea extends inland to meet the mouth of a river. Estuaries are often rich with many different types of plant and animals species, because the freshwater in these areas usually has a high concentration of nutrients and sediments. The waters in estuaries are usually quite shallow, which means that the water is fairly warm and that plants and animals in these locations can receive significant amounts of sunlight. Some subcategories of estuarine environments that you should know for the exam are salt water marshes, mangrove forests, inlets, bays, and river mouths.

Some of Earth's most important ecologically diverse ecosystems are the areas along the shores of fresh bodies of water known as **wetlands.** Types of wetlands include marshes, swamps, bogs, prairie potholes (which exist seasonally), and flood plains (which occur when excess water flows out of the banks of a river and into a flat valley). So, those are the main types of freshwater bodies you'll need to know. Let's review the stratification of freshwater bodies, get through oceans, and move on to the fun stuff—the impacts of water use on humans.

Vertical Stratification in Freshwater Biomes

In all natural bodies of water, there exist layers of water that vary significantly in their temperature, oxygen content, and nutrient levels. These layers are affected differently by seasonal changes and other disturbances, and this also contributes to how they are categorized.

In freshwater, the layers are the **epilimnion**, which is the uppermost, and thus the most oxygenated, layer; and the **hypolimnion,** which is the lower, colder, and denser layer. The demarcation line between these two layers, at which the temperature shifts dramatically, is the **thermocline.**

These layers are also often delineated based on the types of organisms that can live in them. You should definitely be familiar with the following terms for the AP Environmental Science test, so take note!

- **Littoral zone:** Begins with the very shallow water at the shoreline. Plants and animals that reside in the littoral zone receive abundant sunlight. The end of this zone is defined as the depth at which rooted plants stop growing.

- **Limnetic zone:** Surface of open water; the region that extends to the depth that sunlight can penetrate. Organisms that are residents in this zone are short-lived and rely on sunlight to carry out photosynthesis.

- **Profundal zone:** Water that is too deep for sunlight to penetrate. Because the profundal zone is an aphotic zone (a zone that light cannot reach), photosynthesizing plants or animals cannot live in this region.

- **Benthic zone:** The deepest layer in a body of water; characterized by very low temperatures and low oxygen levels.

Freshwater Lake Zones

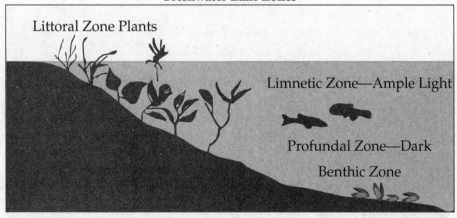

Littoral Zone Plants

Limnetic Zone—Ample Light

Profundal Zone—Dark

Benthic Zone

The World's Oceans

Before we get into our review of the world's oceans, let's consider another aquatic ecosystem (besides wetlands and estuaries) that's an important source of biodiversity; this one is a saltwater ecosystem. Certain landforms that lie off coastal shores are known as **barrier islands.** Because barrier islands are created by the buildup of deposited sediments, their boundaries are constantly shifting as water moves around them. These spits of land are generally the first hit by offshore storms, and they are important buffers for the shoreline behind them.

In tropical waters, a very particular type of barrier island called a **coral reef** is quite common. These barrier islands are formed not from the deposition of sediments, but from a community of living things. The organisms responsible for the creation of coral reefs are cnidarians that secrete a hard, calciferous shell; these shells provide homes and shelter for an incredible diversity of species, but they are also extremely delicate and thus very vulnerable to physical stresses, changes in light intensity, changes in water temperature, and ocean depth and pH.

Like freshwater bodies, oceans are divided into zones based on changes in light and temperature. They're listed below, and again, know them cold for the test!

- **Coastal zone:** This zone consists of the ocean water closest to land. Usually it is defined as being between the shore and the end of the continental shelf.

- **Euphotic zone:** The photic, upper layers of water. The euphotic zone is the warmest region of ocean water; this zone also has the highest levels of dissolved oxygen.

- **Bathyal zone:** The middle region; this zone receives insufficient light for photosynthesis and is colder than the euphotic zone.

- **Abyssal zone:** This is the deepest region of the ocean. This zone is marked by extremely cold temperatures and very low levels of dissolved oxygen, but very high levels of nutrients because of the decaying plant and animal matter that sinks down from the zones above.

Ocean Zones

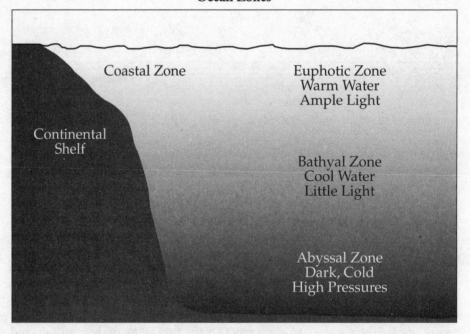

Both freshwater and saltwater bodies experience a seasonal movement of water from the cold and nutrient-rich bottom to the surface. These **upwellings** provide a new nutrient supply for the growth of living organisms in the photic regions. Therefore, they are followed by an almost immediate exponential growth in the population of organisms in these zones, especially the single cell algae, which may form blooms of color called algal blooms. These algae can also produce toxins that may kill fish and poison the beds of filter feeders, such as oysters and mussels. One notorious recurring toxic algal bloom is referred to as **red tide;** this is caused by a proliferation of dinoflagellates.

Water is densest at 4°C. In non-tropical regions of the earth after spring ice melt, the water-surface temperature of lakes and ponds will rise from 0°C to 4°C whereupon this dense surface water will sink to the bottom of the lake or pond. This will displace water at the bottom of the lake or pond to the surface. This overturn brings oxygen to the bottom and nutrients to the top of the lake or pond and occurs during spring and fall as the temperature of the ecosystem changes from cold to warm or the reverse.

Ocean Currents

Ocean currents play a major role in modifying conditions around the earth that can affect where certain climates are located. As the sun warms water in the equatorial regions of the globe, prevailing winds, differences in salinity (saltiness), and Earth's rotation set ocean water in motion. For example, in the Northern Hemisphere, the Gulf Stream carries sun-warmed water along the east coast of the Unites States and as far as Great Britain. This warm water displaces the colder, denser water in the polar regions, which can move south to be re-warmed by the equatorial sun. Northern Europe is kept 5°C to 10°C warmer than it would be were the current not present.

Oceanographers also study a major current, the "ocean conveyer belt" that moves cold water in the depths of the Pacific Ocean while creating major upwellings in other areas of the Pacific.

Ocean Circulation

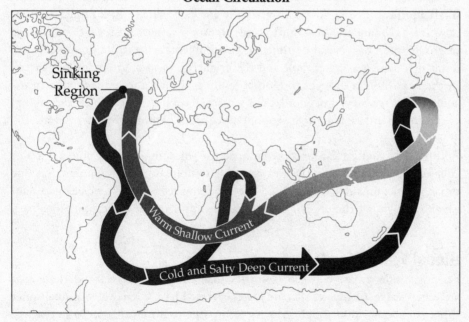

Water, Water, Everywhere...or Not?

As you know, we all need water in order to live. In particular, communities need water for many different industries, including fisheries, recreation, transportation, and agriculture. Agriculture is one of the biggest water-users of all—about 73 percent of the global demand for water is for crop irrigation. Industry accounts for about 21 percent of all water use, and domestic use accounts for about 6 percent.

Since the 1950s, global water use has tripled—mostly due to population growth and improvements in the global standard of living. One way that humans have recently dealt with potential water shortages in communities is through **interbasin transfer**. During interbasin transfer, water is transported very long distances from its source, through aqueducts or pipelines. An example of this type of engineering is the pipeline that now exists between the western and eastern slopes of the Rocky Mountains in Colorado. Known as the Big Thompson Project, 213,000 acre feet of water are delivered annually to the eastern slope of Colorado. However, this method has several negative effects. It can result in different geographic areas arguing over water rights. It can also have serious environmental repercussions; interbasin transfer can increase the salinity of the water and even change the climate of an ecosystem.

In North America especially, humans rely on groundwater as a primary source of water for everyday use. **Groundwater** refers to any water that comes from below the ground; that is, from wells or **aquifers,** which are underground beds or layers of earth, gravel, or porous stone that yield water. Water found in an **unconfined aquifer** is free to flow both vertically and horizontally. A **confined aquifer**, however, has boundaries that don't readily transport water. Our reliance on and use of groundwater has several detrimental environmental effects; for example, it can result in a depressed water table and the drying up of local groundwater sources. In the late 1990s, a drought in Florida resulted in such a severe reduction in the aquifers that roads collapsed for lack of structural support. Subsidence (or sinking) of the Earth's surface is another serious consequence of groundwater withdrawal.

Additionally, aquifers can become compacted—meaning that the mineral grains collapse on each other and the area is unable to hold as much water; and in some urban areas, humans have rendered the groundwater incapable of being replenished by building structures and roads that are impermeable to precipitation.

Global Water Needs

Scientists differentiate between countries that are water-stressed and those that are water-scarce. Countries that are **water-stressed** have a renewable annual water supply of about 1,000–2,000 m^3 per person, but countries that are **water-scarce** have less than 1,000 m^3 per person. Many of the countries that are currently considered water-scarce are developing countries that have rapidly increasing populations—which means that their water scarcity problems will grow over time.

Some of the water-scarce countries that exist today are Algeria, Egypt, Libya, Kenya, Rwanda, Tunisia, Israel, Jordan, Kuwait, Saudi Arabia, Syria, Belgium,

Hungary, the Netherlands, Singapore, Barbados, Malta, Lebanon, Morocco, Niger, Somalia, South Africa, and Sudan. Clearly, that's a lot of countries. Unfortunately, many more are expected to be water-scarce by the year 2050.

Water Use in the United States

The United States is not considered water-scarce, but certain regions of the United States are considered water-stressed. Additionally, water use in the United States is out of control—we use water more quickly than it can possibly be replenished, so water scarcity is definitely in our future if we continue to use water at our present, furious rate.

The hydrologic cycle supplies the water that we use for all of our activities. Water used in our home; manufacturing; cooling equipment that generates electricity; and irrigating croplands are a few examples. The following chart shows the use of freshwater in the United States in 2000, the latest year of available data.

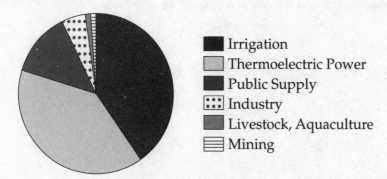

Use	Percentage
Irrigation	40
Thermoelectric Power	39
Public Supply	13
Industry	5
Livestock, Aquaculture	1
Mining	1

What Are We Doing About It?

Water is a tricky business; it's difficult for politicians and lawmakers to put restrictions on water use because many people think that water should be free—after all, it falls from the sky; we can take a bucket from the lake down the street and no one will arrest us for stealing.

For the AP Environmental Science Exam, you should know about certain concepts of human water rights. The first is the idea of riparian right. "Riparian" means "of, on, or relating to the banks of a natural course of water," and **riparian right** is the right of people who have legal rights to use that area. Alternately, in **prior appropriation,** water rights are given to those who have historically used the water in a certain area. In other words, prior appropriation can be thought of as water squatters' rights!

It has been proposed that, in order to solve current global water crises, we simply take the tons of ocean water and desalinate it—this is a fairly simple process physically, but unfortunately it isn't economically viable, because it takes a great deal of energy to remove the salt through distillation or reverse osmosis. As water becomes scarcer globally, it will be important for countries to think of ways to regulate the use of water.

That's everything you'll be expected to know for the exam about water, and you're done with your study of Earth's interdependent systems! Do the questions in the drills found on the next few pages, and move on to the next chapter.

KEY TERMS

Use this list to review the key terms in this chapter. Keep these terms in mind when you are brainstorming hot buttons for your essays!

The Solid Earth

geologic time scale
core
mantle
asthenosphere
lithosphere
crust
tectonic plates
plate boundaries: convergent,
         divergent, transform fault
volcanoes: rift, subduction,
         hot spot
earthquake: focus, epicenter,
         seismograph
subduction
seismograph
tsunami
rock cycle
sedimentary
metamorphic
igneous

Soil

abiotic, biotic
clay, silt, sand
acidity, alkalinity
weathering: physical, chemical,
         biological
soil horizons
humus
leaching
illuviation
arable
loamy
monoculture
crop rotation
green revolution
salinization
land degradation
drip irrigation
erosion
soil laws

Atmosphere

troposphere
greenhouse effect
tropopause
stratosphere
ozone
mesosphere
thermosphere
weather
climate
prevailing winds
Coriolis effect
convection currents
horizontal airflow
dew point
precipitation
convection cell
Hadley cell
trade winds
horse latitudes
doldrums
jet stream
monsoon
rain shadow effect
hurricane
typhoon
El Niño
La Niña
ENSO events

Hydrosphere
watershed
delta
estuary
wetland
epilimnion
hypolimnion
thermocline
freshwater zones: littoral, limnetic,
profundal, benthic
barrier island
saltwater zones: euphotic, bathyal,
abyssal, coastal
upwelling
red tide
interbasin transfer
groundwater
aquifer
unconfined aquifer
confined aquifer
water-stressed, water-scarce
riparian
prior appropriation

Chapter 4 Drill

Directions: Each of the questions or incomplete statements below is followed by five suggested answers or completions. Select the one that is best in each case. For answers and explanations, see Chapter 13.

1. A seismograph is used to measure

   (A) tectonic plate size
   (B) the magnitude of an earthquake
   (C) the temperature of volcanic materials
   (D) the frequency of earthquakes in a region
   (E) soil fertility

2. The aurora borealis occurs in which of the following parts of the atmosphere?

   (A) Troposphere
   (B) Thermosphere
   (C) Mesosphere
   (D) Hydrosphere
   (E) Stratosphere

3. Which of the following are the two most important factors in determining a habitat's climate?

   (A) Temperature and wind speed
   (B) Wind direction and precipitation
   (C) Wind speed and rate of evaporation
   (D) Rate of evaporation and temperature
   (E) Temperature and precipitation

4. The atmosphere is warmed as gases such as water vapor and carbon dioxide absorb the infrared heat radiated from the earth. This process is best described as

   (A) ozone depletion
   (B) the greenhouse effect
   (C) biomagnification
   (D) ionization
   (E) convection

5. The hydrosphere includes all of the following EXCEPT

   (A) watershed
   (B) wetlands
   (C) parent rock
   (D) rivers
   (E) lakes

6. An area where there are cold waters, low oxygen levels, and bottom-dwelling fish. This description best fits the

   (A) benthic zone
   (B) littoral zone
   (C) limnetic zone
   (D) open water zone
   (E) profundal zone

7. The amount of the earth's surface that is covered by water is approximately

   (A) 12 percent
   (B) 36 percent
   (C) 50 percent
   (D) 75 percent
   (E) 93 percent

8. An area where salt and freshwater mix that has a very high level of productivity is correctly called

   (A) the open ocean
   (B) the abyssal zone
   (C) the headwaters
   (D) an estuary
   (E) the littoral zone

9. Which of the following correctly describes the waters in an upwelling area?

   (A) Cold and nutrient rich
   (B) Warm and nutrient poor
   (C) Cold and nutrient poor
   (D) Heavily polluted by human waste
   (E) Shallow and full of light

10. Which of the following best describes an unconfined aquifer? It is an area where

    (A) water always comes to the surface
    (B) water is free to flow in all directions
    (C) water is held in place by impenetrable rocks
    (D) pollutants enter the aquifer
    (E) an aquifer's discharge area is located

11. Which of the following organisms is not likely to be found as a member of the detritus food web?

    (A) Ants
    (B) Earthworms
    (C) Fungi
    (D) Deer
    (E) Bacteria

Directions: Each set of lettered choices below refers to the numbered questions or statements immediately following it. Select the one lettered choice that best answers each question or best fits each statement and then fill in the corresponding oval on the answer sheet. A choice may be used once, more than once, or not at all in each set.

Questions 12-16 deal with the atmosphere.
  (A) atmosphere
  (B) troposphere
  (C) wind
  (D) convection currents
  (E) stratosphere

12. Ozone in this layer blocks UV light from the sun

13. The location of our daily weather

14. The vertical heating and cooling of air

15. The movement of air between masses with different pressures

16. The collection of all gases held to the earth by gravity

Free-Response Question

1. Scientists designed an experiment to learn about the functioning of the hydrologic cycle and the phosphorus cycle in a forest. Using two areas of the same size and geologic features, they cut all the trees down from one plot and did not disturb the other plot. They were able to accurately measure the amount of water that flowed out of the two plots as well as measure the amounts of phosphorus found in the runoff.

   (a) Describe what the differences would be in the volume of water running off the two plots and give one reason why. Assume that the two areas received the same amounts of precipitation.

   (b) Describe the differences in the levels of phosphate found in the runoff of the two plots. Assume that both plots started off with the same amount of phosphorus in the soil.

   (c) Describe one negative effect that might occur in a stream that receives the runoff water and sediment.

   (d) When a tropical rain forest is cut down and used as farmland, the fertility of the soil only lasts a few years. Give an explanation as to why there is little organic matter in rain forest soil and what would happen to that material after deforestation.

REFLECT ACTIVITY

Respond to the following questions:

- For which content topics discussed in this chapter do you feel you have achieved sufficient mastery to answer multiple-choice questions correctly?

- For which content topics discussed in this chapter do you feel you have achieved sufficient mastery to identify hot-button topics and successfully articulate them in free-response questions?

- For which content topics discussed in this chapter do you feel you need more work before you can answer multiple-choice questions correctly?

- For which content topics discussed in this chapter do you feel you need more work before you can successfully identify and articulate them in free-response questions?

- What parts of this chapter are you going to re-review?

- Will you seek further help, outside of this book (such as a teacher, tutor, or AP Central), on any of the content in this chapter—and, if so, on what content?

Chapter 5
The Inhabitants of Planet Earth and Their Relationships

According to the College Board, about 10–15 percent of the questions you'll see on the AP Environmental Science Exam can be classified under the broad topic "The Living World." In this chapter, we'll cover everything you'll need to know to tackle these questions.

As you probably learned in your biology class, the nonliving components of Earth are known as its **abiotic** components. These include the atmosphere, hydrosphere, and lithosphere—the things we studied in the last chapter. In this chapter, we'll talk about the **biotic**, or living components, of Earth. We'll start by discussing elements that bridge the gap between the nonliving and the living—water, nitrogen, carbon, and phosphorus—and how they cycle through the environment. We'll then move on to a discussion of what types of ecosystems exist and how they're structured. We'll continue our review by discussing how energy flows through ecosystems, and we'll end the chapter with a review of how ecosystems change. Let's begin!

CYCLES IN NATURE

As you may have learned in your biology class, nutrients such as carbon, oxygen, nitrogen, phosphorus, sulfur, and water all move through the environment in complex cycles known as **biogeochemical cycles**. Well, you'll need to know a bit about these cycles for the exam, so we'll go through each of them here.

As you can probably tell from the collective name of these natural cycles, living organisms, geologic formations, and chemical substances are all involved in these cycles. Keep in mind that when we describe the movement of these inorganic compounds, it's important to understand both the destinations of the compounds and how they move toward their destinations. In other words, you'll need to know that water moves from the atmosphere to the earth's surface through precipitation, either in the form of snow or rainfall.

For the AP Environmental Science test, it won't be enough for you to know that water moves from the atmosphere to the earth. You'll need to know the different ways it has of getting there.

But, let's talk about a few things that all of these cycles have in common before we go into each one in detail. First of all, the term **reservoir** is used to describe a place where a large quantity of a nutrient sits for a long period of time (in the water cycle, the ocean is an example of a reservoir). The opposite of a reservoir is an **exchange pool**, which is a site where a nutrient sits for only a short period of time (in the water cycle, a cloud is an example of an exchange pool). The amount of time a nutrient spends in a reservoir or an exchange pool is called its **residency time**. In the water cycle, water might exist in the form of a cloud for a few days, but it might exist as part of the ocean for a thousand years! Perhaps surprisingly, living organisms can also serve as exchange pools and reservoirs for certain nutrients; we'll delve into more about this later.

The energy that drives these biogeochemical cycles in the biosphere comes primarily from two sources: the sun, and the heat energy from the mantle and core of the earth. The movements of nutrients in all of these cycles may be via abiotic mechanisms, such as wind, or may occur through biotic mechanisms, such as through living organisms (as we mentioned earlier). Another important fact to note is that while the **Law of Conservation of Matter** states that matter can neither be created nor destroyed, nutrients can be rendered unavailable for cycling through certain processes—for example, in some cycles, nutrients may be transported to deep ocean sediments where they are locked away interminably.

Though we won't get into a discussion of trace elements here, you should also know that certain trace elements such as zinc, copper, and iron are necessary in small amounts for living organisms. Trace elements can cycle in conjunction with the major nutrients, but there's still much to be discovered about these elements and their biogeochemical cycles. For this exam, just know that there are certain trace elements required by living things that cycle, along with the major elements, through the biosphere.

Let's start with perhaps the best-known biogeochemical cycle: the water cycle.

The Water Cycle

As you might imagine, the water that exists in the atmosphere is in a gaseous state, and when it condenses from the gaseous state to form a liquid or solid, it becomes dense enough to fall to the earth because of the pull of gravity. This process is formally known as **precipitation.** When precipitation falls onto the earth, it may travel below ground to become **groundwater,** or it may travel across the land's surface as **runoff** and enter a drainage system, such as a stream or river, that will eventually deposit it into a body of water such as a lake or an ocean. Lakes and oceans are reservoirs for water. In certain cold regions of Earth, water may also be trapped on the earth's surface as snow or ice; in these areas, the blocks of snow or ice are reservoirs.

Water is also cycled through living systems. For example, plants consume water (and carbon dioxide) in the process of photosynthesis, in which they produce carbohydrates. Because all living organisms are primarily made up of water, they act as exchange pools for water.

Water is returned to the atmosphere from both the earth's surface and from living organisms in a process called **evaporation.** Specifically, animals respire and release water vapor and additional gases to the atmosphere. In plants, the process of **transpiration** releases large amounts of water into the air. Finally, other major contributors to atmospheric water are the vast number of lakes and oceans on Earth's surface. Incredibly large amounts of water continually evaporate from their surfaces.

Take a look at the graphic on the next page, which shows all of the forms that water takes in the biosphere and atmosphere.

The Water Cycle

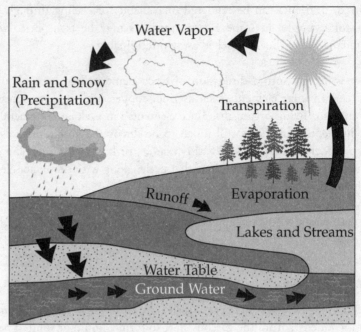

The Carbon Cycle

Now let's talk about carbon. The key events in the carbon cycle are **respiration,** in which animals (and plants!) breathe in oxygen and give off carbon dioxide, and **photosynthesis,** in which plants take in carbon dioxide, water, and energy from the sun to produce carbohydrates. In other words, living things act as exchange pools for carbon.

When plants are eaten by animal consumers, the carbon locked in the plant carbohydrates passes to other organisms and continues through the food chain (more on this later in the chapter). In turn, when organisms—both plants and animals—die, their bodies are decomposed through the actions of bacteria and fungi in the soil; this releases CO_2 back into the atmosphere.

One aspect of the carbon cycle that you should definitely be familiar with for the exam is this: When the bodies of once-living organisms are buried and subjected to conditions of extreme heat and extreme pressure, this organic matter eventually becomes oil, coal, and gas. Oil, coal, and natural gas are collectively known as fossil fuels, and when fossil fuels are burned, or **combusted,** carbon is released into the atmosphere. Finally, carbon is also released into the atmosphere through volcanic action.

There are three major reservoirs of carbon; the first is the world's oceans, because CO_2 is very soluble in water. The second large reservoir of CO_2 is Earth's rocks. Many types of rocks—called carbonate rocks—contain carbon, in the form of calcium carbonate. Finally, fossil fuels are a huge reservoir of carbon, as well.

The Carbon Cycle

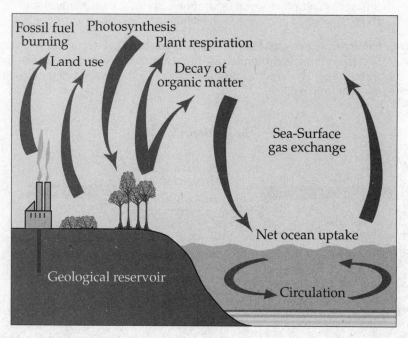

The Nitrogen Cycle

Earth's atmosphere is made up of approximately 78 percent nitrogen and 21 percent oxygen (the other components of the atmosphere are trace elements); nitrogen is the most abundant element in the atmosphere. For this reason, it might not seem like living organisms would find it difficult to get the nitrogen they need in order to live. But it is! This is because atmospheric N_2 is not in a form that can be used directly by most organisms. In order to keep this rather complicated cycle straight, let's look at it in steps.

Step 1: Nitrogen fixation—In order to be used by most living organisms, nitrogen must be present in the form of ammonia (NH_3) or nitrates (NO_3^-). Atmospheric nitrogen can be converted into these forms, or "fixed," by atmospheric effects such as lightning storms, but most nitrogen fixation is the result of the actions of certain soil bacteria. Fixing is the process that allows nitrogen to be made biologically available, much as photosynthesis makes carbon biologically available. One important soil bacteria that participates in nitrogen fixation is *Rhizobium*. These nitrogen-fixing bacteria are often associated with the roots of legumes such as beans or clover. In the future, we may be able to insert the genes for nitrogen fixation into crop plants, such as corn, and reduce the amount of fertilizer that is used.

Step 2: Nitrification—In this process, soil bacteria converts ammonium (NH_4^+) into one of the forms that can be used by plants—nitrate (NO_3).

Step 3: Assimilation—In assimilation, plants absorb ammonium (NH_3), ammonia ions (NH_4^+), and nitrate ions (NO_3^-) through their roots. Heterotrophs, or organisms that receive energy by consuming other organisms, then obtain nitrogen when they consume plants' proteins and nucleic acids.

Step 4: Ammonification—In this process, decomposing bacteria convert dead organisms and other waste to ammonia (NH_3) or ammonium ions (NH_4^+), which can be reused by plants.

Step 5: Denitrification—In denitrification, specialized bacteria (mostly anaerobic bacteria) convert ammonia back into nitrites and nitrates and then into nitrogen gas (N_2) and nitrous oxide gas (N_2O). These gases then rise to the atmosphere.

The Nitrogen Cycle

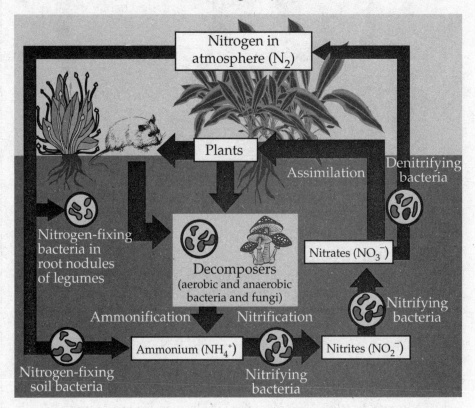

The Phosphorus Cycle

The **phosphorus cycle** is perhaps the simplest biogeochemical cycle, mostly because phosphorus does not exist in the atmosphere outside of dust particles. Phosphorus is necessary for living organisms because it's a major component of nucleic acids and other important biological molecules. One important idea for you to remember about the phosphorus cycle is that phosphorus cycles are more local than those of the other important biological compounds.

For the most part, phosphorus is found in soil, rock, and sediments; it's released from these rock forms through the process of chemical weathering. Phosphorus is usually released in the form of phosphate (PO_4^{3-}), which is soluble and can be absorbed from the soil by plants. You should know that phosphorus is often a limiting factor for plant growth, so plants that have little phosphorus are stunted.

Phosphates that enter the water table and travel to the oceans can eventually be incorporated into rocks in the ocean floor. Through geologic processes, ocean mixing, and upwelling, these rocks from the seafloor may rise up so that their components once again enter the **terrestrial cycle.** Take a look at the phosphorus cycle shown in the diagram below.

Humans have affected the phosphorus cycle by mining phosphorus-rich rocks in order to produce fertilizers. The fertilizers placed on fields can easily leach into the groundwater and find their way into aquatic ecosystems where they can cause eutrophication. **Eutrophication** occurs when a body of water receives excess nutrients. The adundance of nutrients can cause an overgrowth of algae and deplete the water of oxygen.

The Phosphorous Cycle

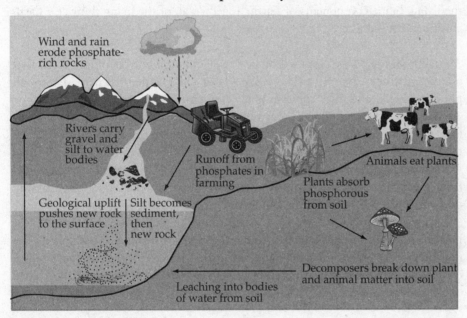

We are almost done with the chemistry. But we need to discuss one more element before we move on to discuss the biosphere, and that's sulfur.

Sulfur

The last biogeochemical cycle we'll talk about is the sulfur cycle. Sulfur is one of the components that make up proteins and vitamins, so plants and animals both need sulfur in their diets. Plants absorb sulfur when it is dissolved in water, so they can take it up through their roots when it's dissolved in groundwater. Animals obtain sulfur by consuming plants.

Most of the earth's sulfur is tied up in rocks and salts or buried deep in the ocean in oceanic sediments, but some sulfur can be found in the atmosphere. The natural ways that sulfur enters the atmosphere are through volcanic eruptions, certain bacterial functions, and the decay of once-living organisms. When sulfur enters the atmosphere through human activity, it's mainly via industrial processes that produce sulfur dioxide (SO_2) and hydrogen sulfide (H_2S) gases. We'll talk more about sulfur and how it contributes to air pollution in Chapter 9.

All right! It's time to move on to our discussion of the biotic components of Earth. Let's start with a review of how energy moves through ecosystems.

FOOD CHAINS AND FOOD WEBS

You might recall that the nonliving components of the environment are known as the abiotic components. These include the atmosphere, hydrosphere, and lithosphere. Remember them from Chapter 4? Well, it's time to begin our study of the living, biotic components of Earth. Together, all of the living things on Earth constitute the biosphere.

All living things can be classified by how they obtain food. You might recall that plants and some caynobacteria are capable of making their own food through photosynthesis, and that some animals (for example, mice) eat plants. Some animals (for example, humans) eat both plants and animals, and some animals (for example, wolves) eat only other animals. There are actually two fancy terms that are normally used to describe these broad categories of organisms: **Autotrophs** are those organisms that can produce their own organic compounds from inorganic chemicals, while **heterotrophs** obtain food energy by consuming other organisms or products created by other organisms.

Finally, as unpleasant as it might be to think about, some animals feed only on the remains of other plants and animals! All of these different types of living things fall into specific categories—and you will definitely need to memorize all of these terms before the test, if you don't already know them!

Producers

Producers are organisms that are capable of converting radiant energy, or chemical energy, into carbohydrates. The group of producers includes plants and algae, both of which can carry out photosynthesis. The unbalanced overall reaction of photosynthesis is shown below.

$$H_2O + CO_2 + \text{solar energy} \rightarrow CH_2O + O_2$$

While most producers make food through photosynthesis, a few autotrophs make food from inorganic chemicals in **anaerobic** (without oxygen) environments, through the process of chemosynthesis. Chemosynthesis is only carried out by a few specialized bacteria, called **chemotrophs,** some of which are found in hydrothermal vents deep in the ocean. This unbalanced reaction is shown below.

$$O_2 + H_2S + O_2 + \text{energy} \rightarrow CH_2O + S + H_2O$$

At this point, let's discuss a few other environmental science terms that you'll be required to know for the exam. The **Net Primary Productivity** (**NPP**) is the amount of energy that plants pass on to the community of herbivores in an ecosystem. It is calculated by taking the **Gross Primary Productivity,** which is the amount of sugar that the plants produce in photosynthesis, and subtracting from it the amount of energy the plants need for growth, maintenance, repair, and reproduction. NPP is measured in kilocalories per square meter per year ($kcal/m^2/y$). In other words, the gross primary productivity of an ecosystem is the rate at which the producers are converting solar energy to chemical energy (or, in a hydrothermal ecosystem, the rate of productivity of the chemotrophs). Perhaps not surprisingly, the net productivity of an ecosystem is a limiting factor for its number of consumers. A limiting factor is a factor that control's a population's growth. It can be many things: space, available food, water, and as we just mentioned, the net productivity of an ecosystem.

Consumers

Consumers are organisms that must obtain food energy from secondary sources, for example, by eating plant or animal matter. There are a number of different types of consumers, and we've listed them below. Commit these to memory!

- **Primary consumers:** This category includes the herbivores, which consume only producers (plants and algae).

- **Secondary consumers:** An organism that consumes a primary consumer is as secondary consumer.

- **Tertiary consumers:** An organism that consumes a secondary consumer is a tertiary consumer.

- **Detritivores:** The organisms in this group derive energy from consuming nonliving organic matter such as dead animals or fallen leaves.

- **Decomposers:** These are bacteria or fungi that absorb nutrients from nonliving organic matter such as plant material, the wastes of living organisms, and corpses. They convert these materials into inorganic forms.

Note that one organism may occupy multiple levels of a food chain. By eating a hamburger with toppings you are both a primary consumer because you are eating tomatoes and lettuce, and a secondary consumer by eating the beef.

Let's move on and talk about how energy flows through all of these different types of organisms in ecosystems.

Food Chains

As you probably recall, energy flows in one direction through ecosystems: from the sun to producers, to primary consumers, to secondary consumers, to tertiary consumers. In an ecosystem, each of these feeding levels is referred to as a **trophic level**. With each successive trophic level, the amount of energy that's available to the next level decreases. In fact, only about 10 percent of the energy from one trophic level is passed to the next; most is lost as heat, and some is used for metabolism and anabolism. Interestingly enough, this is why food chains rarely have more than four trophic levels.

Food chains are usually represented as a series of steps, in which the bottom step is the producer and the top step is a secondary or tertiary consumer. In food chains, the arrows depict the transfer of energy through the levels, and in fancier food chains, the relative biomass (the dry weight of the group of organisms) of each trophic level will often be represented. Here's a simple food chain.

Food Chain

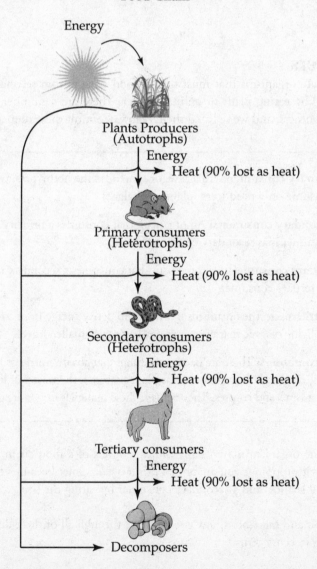

What we're showing here is a typical terrestrial food chain, but keep in mind that there are aquatic food chains as well, with algae and different types of fish.

One final note about food chains: In a food chain, only about 10 percent of the energy is transferred from one level to the next. The other 90 percent is used for things like respiration, digestion, running away from predators—that is, it's used to power the organism doing the eating! In other words, the producers have the most energy in an ecosystem; the primary consumers have less energy than producers; the secondary consumers have less energy than the primary consumers; and the tertiary consumers will have the least energy of all. The amount of energy (in kilo-calories) available at each trophic level organized from greatest to least is an **energy pyramid,** as seen on the next page.

Energy Pyramid

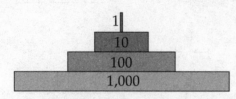

Tertiary consumers
Secondary consumers
Primary consumers
Producers

Hypothetical units of energy

Note that 90% of energy is lost
at each step

Biomagnification

Food chains represent the flow of energy in an ecosystem, but other things can flow through food chains, too—including environmental toxins. While the amount of useable energy decreases at every level of the food chain, the concentration of certain toxins increases at each successive level, since most toxins cannot be broken down by organisms. The term **bioaccumulation** is used to describe the accumulation of a substance, such as a toxic chemical, in the tissues of a living organism, such as a producer or a primary consumer.

Biomagnification is the term used to describe the increasing concentration of these toxin molecules at successively higher trophic levels in a food chain. Keep in mind that although really any type of molecule could be described using the terms bioaccumulation and biomagnification, generally these terms are used to describe toxins and heavy metals. And certainly, this is how they'll be used on the test.

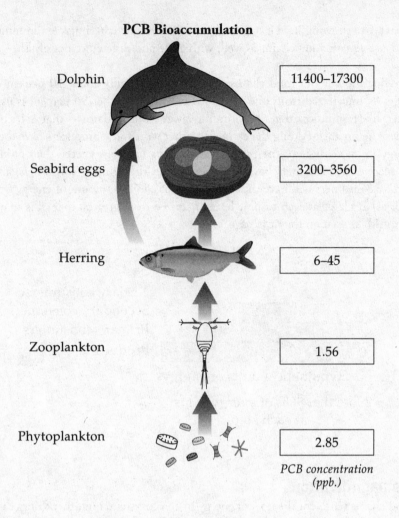

PCB Bioaccumulation

Dolphin	11400–17300
Seabird eggs	3200–3560
Herring	6–45
Zooplankton	1.56
Phytoplankton	2.85

PCB concentration (ppb.)

Food Webs—Tangled Food Chains

As you're probably already aware, food chains are an oversimplified way of demonstrating the myriad of feeding relationships that exist in ecosystems. Because there are so many different types of species of plants and animals in ecosystems, their relationships in real-world ecosystems are much more complicated than can be depicted in a single food chain. Therefore, we use a **food web** in order to represent feeding relationships in ecosystems more realistically.

Food Web

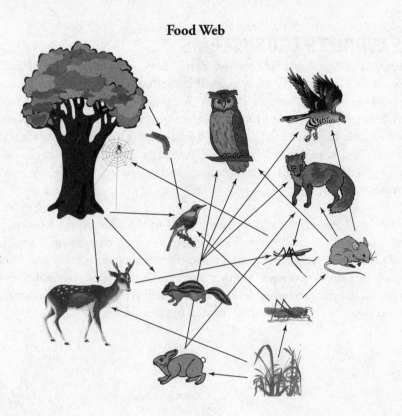

Again, this is a typical terrestrial food web, but keep in mind that very complicated aquatic food webs exist as well! Let's take a step back for a minute and discuss the setting for food chains and food webs—ecosystems.

THE WORLD'S ECOSYSTEMS

Because different geographic areas on Earth differ so much in their abiotic and biotic components, we can easily place them in broad categories. The two largest categories are broken down in this way: Ecosystems that are based on land are called **biomes,** while those in aqueous environments are known as **aquatic life zones.** Aquatic ecosystems are categorized primarily by the salinity of their water—freshwater and saltwater ecosystems fall into separate categories. Land environments are separated into biomes based on their climate (remember temperature and precipitation!).

Although it might seem that each biome listed in the table on the following page is very distinct, in reality, biomes blend into each other; they do not have distinct boundaries. The transitional area where two ecosystems meet actually has a name—these areas are called **ecotones.** Another important term that you should be familiar with for the exam is **ecozones:** Ecozones (also called **ecoregions**) are smaller regions within ecosystems that share similar physical features. Here's that table.

Types of Ecosystems			
Biome	**Annual Rainfall, Soil Type**	**Major Vegetation**	**World Location**
Deciduous forest (temperate and tropical)	75–250 cm, rich soil with high organic content	Hardwood trees	North America, Europe, Australia, and Eastern Asia
Tropical rainforest	200–400 cm, poor quality soil	Tall trees with few lower limbs, vines, epiphytes, plants adapted to low light intensity	South America, West Africa, and Southeast Asia
Grasslands	10–60 cm, rich soil	Sod-forming grasses	North American plains and prairies; Russian steppes; South African velds; Argentinean pampas
Coniferous forest (Taiga)	20–60 cm— mostly in summer, soil is acidic due to vegetation	Coniferous trees	Northern North America, northern Eurasia
Tundra	Less than 25 cm, soil is permafrost	Herbaceous plants	The northern latitudes of North America, Europe, and Russia
Chaparral (scrub forest)	50–75 cm— mostly in winter, soil is shallow and infertile	Small trees with large hard leaves, spiny shrubs	Western North America, the Mediterranean region
Deserts (cold and hot)	Less than 25 cm, soil has a course texture (i.e., sandy)	Cactus, other low-water adapted plants	30 degrees north and south of the equator

Not surprisingly, each biome has specific characteristics that determine the types of organisms that are capable of living in it. Some of these characteristics are the type and availability of nutrients, the ecosystems' temperature, the availability of water, and how much sunlight the region receives. One important law to be familiar with for this test is the **Law of Tolerance.** The Law of Tolerance describes the degree to which living organisms are capable of tolerating changes in their environment. Living organisms exhibit a range of tolerance, and even individuals within a population tolerate changes to their environment differently: This concept is the basis for natural selection, which drives evolution (more on this later in the chapter).

Another important law for you to know is the **Law of the Minimum,** which states that living organisms will continue to live, consuming available materials until the supply of these materials is exhausted.

Ecosystem Diversity

The term biodiversity is used to describe the number and variety of organisms found within a specified geographic region, or ecosystem. It also refers to the variability among living organisms, including the variability within and between species and within and between ecosystems. Therefore, when we talk about the biodiversity of an area, we must specify the aspect of biodiversity that we're describing, or else the term is too vague to be comprehensible. In general, however, biodiversity in an ecosystem is a good thing. The more biodiversity in a certain species within an ecosystem, the larger and more diverse the species' gene pool, and the greater its chance of adaptation, and thus survival.

EVOLUTION

Biodiversity in all forms is the result of **evolution.** Evolution is the change in a population's genetic composition over time.

We use a figure called a **phylogenetic tree** to model evolution. Phylogenetic trees can be very broad, like the one on the next page, which encompasses many types of species, or they can be very specific, and describe the evolutionary relationships that exist between two species (or even the genomes of one species!).

While you won't need to know much about evolution for this exam, you will need to have a rough idea of how and why it takes place, so we'll run through that now. Without trying to recreate the evolution of all living organisms, we will limit our discussion to a description of how new species are formed. This process is called **speciation**.

Strictly speaking, a **species** is defined as a group of organisms that are capable of breeding with one another—and incapable of breeding with other species. As you may recall, individual organisms that are better adapted for their environment will live and reproduce, ensuring that their genes are part of their population's next generation. This is what Charles Darwin meant by **evolutionary fitness.**

Phylogenetic Tree

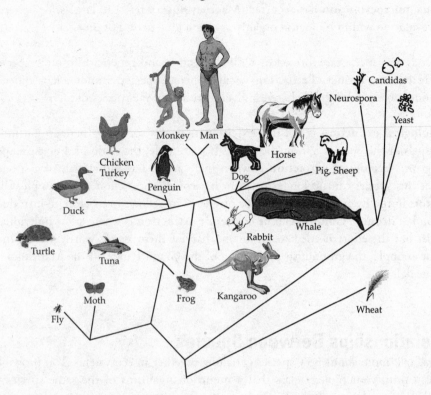

How Evolution Works

When a habitat (an organism's physical surroundings) selects certain organisms to live and reproduce and others to die, that population is said to be undergoing **natural selection**. In natural selection, beneficial characteristics that can be inherited are passed down to the next generation, and unfavorable characteristics that can be inherited become less common in the population. It is important to remember that natural selection acts upon a whole population, not on an individual organism during its lifetime. What changes during evolution is the total genetic makeup of the population, or **gene pool,** and natural selection is one of the mechanisms by which evolution operates.

The other way evolution operates is **genetic drift**. Genetic drift is the accumulation of changes in the frequency of alleles (versions of a gene) over time due to sampling errors—changes that occur as a result of random chance. For example, in a population of owls there may be an equal chance of a newly born owlet having long talons or short talons, but due to random breeding variances a slightly larger number of long-taloned owlets are born. Over many generations, this slight variance can develop into a larger trend, until the majority of owls in that population have long talons. These breeding variances could be a result of a chance event—such as an earthquake that drastically reduces the size of the nesting population one year. Small populations are more sensitive to the effects of genetic drift than large, diverse populations.

When a population displays small scale changes over a relatively short period of time, **microevolution** has occurred. **Macroevolution** refers to large-scale patterns of evolution within biological organisms over a long period of time.

Just as new species are formed by natural selection and genetic drift, other species may become extinct. **Extinction** occurs when a species cannot adapt quickly enough to environmental change and all members of the species die.

Biological extinction is the true extermination of a species. There are no individuals of this species left on the planet (for example, the dodo bird or passenger pigeon). **Ecological extinction** is when there are so few individuals of a species that this species can no longer perform its ecological function (for example, alligators in the Everglades in the 1960s or wolves in Yellowstone before re-introduction last decade). **Commercial or economic extinction** is where a few individuals exist but the effort needed to locate and harvest them is not worth the expense (for example, the groundfish population of the Grand Banks off the Maritimes of Canada).

Relationships Between Species

Let's talk more about how species get along together in ecosystems. You probably recall from your biology class that a group of organisms of the same species is called a **population,** and when populations of different species occupy the same geographic area, they form a **community.** Every species within a community has an ecological niche. A species' **niche** is described as the total sum of a species' use of the biotic and abiotic resources in its environment. The niche describes where the species lives, what it eats, and all of the other resources the species utilizes in an ecosystem. Another term you should know for the exam is habitat—a **habitat** is the area or environment where an organism or ecological community normally lives or occurs.

Some species interact quite a bit with other members of their population; for example, some animals form herds, while other species are loners—like bears. The reasons for these different levels of sociability are largely competition, predation, and a general need to exploit the resources in the environment.

Competition arises when two individuals—of the same species or of different species—are competing for resources in the environment. When the two individuals that are competing are of the same species, this is called **intraspecific competition,** and when they are of different species, it's called **interspecific competition.** The resources that are competed for can be food, air, shelter, sunlight, and various other factors necessary for life; individuals may be competing to live in a fallen tree, to catch a running rabbit, or to mate with the most desirable female in the population. The competitor who is "most fit" eventually wins and obtains the resource. That's right—the others are eliminated by competition.

One more thing about competition: When two different species in a region compete, and the better adapted species wins, this phenomenon is called **competitive exclusion. Gause's principle** states that no two species can occupy the same niche at the same time, and that the species that is less fit to live in the environment will relocate, die out, or occupy a smaller niche. When a species occupies a smaller niche than it would in the absence of competition, this compromised niche is called its **realized niche.** (The niche it would have if there were no competition is known as its **fundamental niche.**) Direct competition can also be avoided in the cause of **resource partitioning**. This occurs when different species use slightly different parts of the habitat, but rely on the same resource. For example, there are five species of warblers that can all live in the same pine tree. They can coexist because each species feeds in a different part of the tree: the trunk, at the ends of the branches, and at other sites. All right, moving on!

Know the three types of relationships between species: competition, predation, and symbiotic relationships.

Although it's relatively easy to observe competition between animals, competition between plants is much more subtle and occurs much more slowly. However, if you have a few years to kill, spend some time in your backyard watching the trees and other plants grow. You'll see that they compete for sunlight and for ground space; they even produce chemicals that inhibit other plants' growth!

The second important type of interspecies interaction is predation. **Predation** occurs when one species feeds on another, and it is the force that drives changes in population size. For example, in a year in which rainfall is relatively high in some regions, rabbits have plenty of food; this enables them to reproduce very successfully, and the number of rabbits in a population will increase dramatically. In turn, if the coyote is a predator of the rabbit, coyotes will have plenty of food, and their population will also boom. However, if the following year the rainfall is below average, there will be less grass; the population of rabbits will decline, and this will result in a decline in the population of coyotes. As a final note about predation: While it's tempting to think of predation existing only between animals, remember that herbivores prey on plants and zooplankton on phytoplankton!

A third type of relationship that exists between organisms is the symbiotic relationship. **Symbiotic relationships** are close, prolonged associations between two or more different organisms of different species that may, but do not necessarily, benefit each member. There are three types of symbiotic relationships, and you should be familiar with all of these for exam day. In mutualistic symbiotic relationships **(mutualism)** both species benefit; for example, this type of relationship exists between sea anemone and clown fish. The clown fish protects the sea anemone from some of its predators, while the stinging cells of the anemone protect the clown fish; the fish also eats some of the detritus left behind when the anemone feeds. In commensalistic symbiotic relationships **(commensalism)**, one organism benefits while the other is neither helped nor hurt. One example of this type of relationship exists between trees and epiphytes (bromeliads and some orchids). The trees are not affected by the epiphytes growing in them, and the epiphytes benefit by collecting water running down the bark and get better access to light than they would on the ground. Finally, **parasitism** is a relationship in which one species is harmed and the other benefits; for example, the relationship that exists between fleas and dogs.

HOW ECOSYSTEMS CHANGE

Believe it or not, oftentimes the biotic balance in a community is maintained by a single species, known as the **keystone species.** The name "keystone" comes from the last stone placed in an arch bridge, which is the key stone. A keystone species is a species whose very presence contributes to an ecosystem's diversity and whose extinction would consequently lead to the extinction of other forms of life. For example, fig trees are the keystone species in a tropical forest; likewise wolves were introduced back into Yellowstone Park because without wolves to control the number of herbivores, the ecosystem had drastically changed. As a general rule, if the keystone species is removed from an ecosystem, then the ecosystem completely changes.

Indicator species are species that are used as a standard to evaluate the health of an ecosystem. They are more sensitive to biological changes within their ecosystems than are other species, so they can be used as an early warning system to detect dangerous changes to a community. Trout are a common indicator species, because they are particularly sensitive to pollutants in water. The disappearance of trout from a particular habitat is a warning that that habitat is becoming polluted.

Indigenous species are those that originate and live or occur naturally in an area or environment. With increasing frequency, however, new species are being introduced into ecosystems by chance, by accident, or with intention. While some introduced species cannot find a niche and die out, many others are quite happy in their new environment, and compete successfully with the indigenous species. One example of this is grey squirrels, which were introduced to England in 1876. The grey squirrel competed with England's native species of squirrel, the red squirrel, and today there are fewer than 30,000 red squirrels alive in England. Another example of the harm that introduced species can do was seen when, in 1904, a fungus was introduced accidentally into the deciduous forests of the eastern United States; this fungus caused a blight that killed nearly all of the chestnut trees by the early 1950s.

Although some people don't like to use the term invasive species because they feel that it's derogatory, it is often used to describe introduced species. Two other examples of invasive species are zebra mussels, which were introduced into the Great Lakes when ships dumped ballast water into the lakes, and the quickly growing vine kudzu, which was originally introduced in the southeastern United States in order to control the problem of erosion.

Ecological Succession

Communities are not static; they are constantly changing. Species of plants and animals are continually coming and going; evolving and dying out. Some of the changes that take place in a geographic area are predictable ones that can be described as **ecological succession.**

If ecological succession begins in a virtually lifeless area, such as the area below a retreating glacier, it is called **primary succession. Secondary succession** is ecological succession that takes place where an existing community has been cleared (by events such as fire, tornado, or human impact), but the soil has been left intact. The organisms in the first stages of either type of succession are referred to as **pioneer species,** and typically have wide ranges of environmental tolerance. The communities in each stage of succession drive the environmental changes that will allow the next stage to take over. The final stage of succession, in which there is a dynamic balance between the abiotic and biotic components of the community, is referred to as the **climax community.**

How does a new habitat full of bare rocks eventually turn into a forest? The first stage of the job usually falls to a community of lichens. Lichens are hardy organisms. They can invade an area, land on bare rocks and erode the rock surface, and over time turn them into soil. Lichens are pioneer organisms. Once lichens have made an area more habitable, other organisms can settle in. Lichens are replaced (out-competed!) by mosses and ferns, which in turn are replaced by tough grasses, then low shrubs, then conifers, then short-lived hardwood trees such as dogwood and red maple trees, and finally long-lived hardwood trees. Refer to the flowchart of ecological succession for a deciduous forest. Note that the stages are classified by the major new plant group, but remember that with the introduction of each new plant species comes an array of different animal species that prey upon it.

We're almost done with this chapter, but you need to know a few more terms and concepts before you can move on to the population chapter. The following material will almost certainly be asked about on the test.

When the size of an organism's natural habitat is reduced, or when, for example, development occurs that isolates the habitat, this process is called **habitat fragmentation.** Habitat fragmentation can be quite damaging. As you know, ecosystems are not isolated; they abut each other and meet at wide and overlapping boundaries, called **ecotones.** At these boundaries, there is greater species diversity and biological density than there

Ecological Succession

Bare rock
↓
Lichen, Algae, Mosses, Bacteria
(Break down rock and leave organic debris
which together forms soil)
↓
Grasses
(Add organic matter to soil and anchor it in place)
↓
Small herbaceous plants
(Continue to add organic matter to soil)
↓
Small bushes
(Add shelter and shade for other plants)
↓
Conifers
(Create additional habitats)
↓
Short-lived hardwoods such as
dogwood and red maple
(Can tolerate shade of conifers but
are short-lived and vulnerable to damage)
↓
Long-lived hardwoods
(More specialized, hardier hardwoods
such as oak and hickory)

is in the heart of ecological communities, and this is called the **edge effect.** Some species can only live on the edge of certain habitats, and if the boundaries of a habitat are changed, a new edge is created, and both the edge and interior habitats are damaged.

The **theory of island biogeography** was first developed to address species diversity on actual islands, but has since been extended to apply to habitable areas that are surrounded by unhabitable areas. According to the theory, the number of species found on an undisturbed "island" is determined by two factors: immigration and extinction.

Whew. You're done with this chapter! Congratulations. Before moving on to the next chapter (Population Ecology), do the questions in the drill—and don't forget to use the techniques you learned in Part III.

KEY TERMS

Make sure you know these words and how to use them in your essays!

Cycles in Nature

biotic
abiotic
biogeochemical cycles
reservoir
exchange pool
residency time
Law of Conservation of Matter
precipitation
groundwater
runoff
evaporation
transpiration
respiration
photosynthesis
combusted
nitrogen fixation
nitrification
assimilation
ammonification
denitrification
phosphorus cycle
terrestrial cycle
eutrophication
sulfur cycle

Food Chains and Food Webs

autotroph
heterotroph
chemotroph
producer
Net Primary Productivity & gross
        primary productivity
consumers: primary, secondary,
        tertiary
detrivore
decomposer
trophic level
food chain
energy pyramid
bioaccumulation
biomagnification
food web

Ecosystems

biomes
aquatic life zones
ecotone
ecozone
deciduous forests
coniferous forests
tropical rainforest
grasslands
tundra
chaparral
deserts
Law of Tolerance
Law of the Minimum

Evolution

evolution
phylogenetic tree
speciation
competitive exclusion
resource partitioning
microevolution
macroevolution
population
community
niche
habitat
competition: intraspecific, interspecific
Gause's principle
realized niche, fundamental niche
predation
symbiotic relationship
mutualism
commensalism
parasitism
species
natural selection
gene pool
genetic drift
extinction
biological extinction
ecological extinction
commercial/economic extinction
evolutionary fitness

How Ecosystems Change
keystone species
indicator species
indigenous species
invasive species
ecological succession
primary succession
secondary succession
pioneer species
climax community
habitat fragmentation
edge effect
island biography theory

Chapter 5 Drill

Directions: Each of the questions or incomplete statements below is followed by five suggested answers or completions. Select the one that is best in each case. For answers and explanations, see Chapter 13.

1. The relationship between a tick and a bird is best described as which of the following?

    (A) Commensalism
    (B) Mutualism
    (C) Parasitism
    (D) Neutralism
    (E) Competition

2. When two species live in the same habitat and use exactly the same resources, which of the following will probably occur?

    (A) The two species can live together indefinitely.
    (B) One of the species will eventually go extinct.
    (C) One species will evolve into a parasite.
    (D) The two species do not interact.
    (E) This competition does not occur in nature.

3. Organisms use different resources in the same habitat, and in this way avoid competition. This is referred to as

    (A) the Law of Tolerance
    (B) hunting and gathering
    (C) predator-prey relationship
    (D) resource partitioning
    (E) commensalism

4. Which of the following is true about the roles of both parasites and predators in ecosystems?

    (A) Predators and parasites can act as environmental resistance and allow the host population to grow.
    (B) Predators are generally smaller and parasites support
    many predators.
    (C) Predators generally have specialized means to capture
    prey.
    (D) Predators and parasites can divide the host population
    so that both can feed off the hosts.
    (E) Parasites and predators eliminate the weak and sick,
    leaving the strongest to reproduce.

5. All of the following are true concerning the characteristics of a climax community EXCEPT

    (A) the adult plants are small in size
    (B) there are many different species of plants
    (C) there is a mixture of decomposers, producers, and consumers
    (D) most of the organisms are specialists in their niche requirements
    (E) there is a large amount of biomass

6. Which of the following describes the direction of the flow of energy in a food chain?

    (A) From parasite to host
    (B) From predator to prey
    (C) From prey to predator
    (D) From one mutual to another
    (E) From prey to commensal

7. Which of the following element's cycles includes long-term storage in rocks and a short storage time in the atmosphere?

    (A) Sulfur
    (B) Carbon
    (C) Nitrogen
    (D) Calcium
    (E) Uranium

8. The current trend where some species of bacteria have become resistant to antibiotics is best described as

    (A) genetic diversity
    (B) speciation
    (C) extinction
    (D) macroevolution
    (E) microevolution

9. Large herds of grazing mammals are most likely to be located in a

    (A) rain forest
    (B) estuary
    (C) coniferous forest
    (D) grasslands
    (E) desert

Directions: Each set of lettered choices below refers to the numbered questions or statements immediately following it. Select the one lettered choice that best answers each question or best fits each statement. A choice may be used once, more than once, or not at all in each set.

Questions 10-14 refer to the process of succession.
    (A) Inertia
    (B) Disturbance
    (C) Primary succession
    (D) Secondary succession
    (E) Tolerance

10. When late succession plants are not disturbed by early succession plants

11. When succession starts from an area where humans once farmed

12. When a community starts from bare rock

13. The tendency of an ecosystem to maintain its overall structure

14. An event that will instigate the process of succession

Questions 15-19 deal with types of species.
    (A) specialist species
    (B) keystone species
    (C) native species
    (D) alien species
    (E) indicator species

15. The species that normally live and thrive in a habitat

16. Species that play a pivotal role in the habitat

17. A species whose decline indicates damage to the habitat

18. A nonnative species

19. A species with a narrow niche, which can only live in a certain habitat

Free-Response Question

1. Students from a local high school participated in a study of Hillside Pond. After safely taking samples of some small fish, a fish-eating hawk, some pond water, some zooplankton, and a fish that preys on the small fish, they determined the average concentration of compound "X" in each sample. The table below summarizes their data.

Organism	Compound "X" concentration
Small fish	0.1 ppm
Hawk	3.0 ppm
Pond water	0.1 ppb
Zooplankton	0.2 ppb
Predatory fish	1.0 ppm

(a) Describe one process that would cause compound "X" to contaminate the pond's water.

(b) Draw a food chain that illustrates the correct trophic order in the pond. Include the concentrations of compound "X" for each part of the chain.

(c) Describe a process that would explain the different concentrations of compound "X" in each organism.

(d) Describe one real-life example of a substance that behaves like compound "X" in the oceans. Give one negative effect that the substance might have on humans.

REFLECT ACTIVITY

Respond to the following questions:

- For which content topics discussed in this chapter do you feel you have achieved sufficient mastery to answer multiple-choice questions correctly?

- For which content topics discussed in this chapter do you feel you have achieved sufficient mastery to identify hot-button topics and successfully articulate them in free-response questions?

- For which content topics discussed in this chapter do you feel you need more work before you can answer multiple-choice questions correctly?

- For which content topics discussed in this chapter do you feel you need more work before you can successfully identify and articulate them in free-response questions?

- What parts of this chapter are you going to re-review?

- Will you seek further help, outside of this book (such as a teacher, tutor, or AP Central), on any of the content in this chapter—and, if so, on what content?

Chapter 6
Population
Ecology

As we mentioned on page 14, about 10–15 percent of the exam will be about population. In other words, you'll definitely see some questions on the topics in this chapter, so learn the bold terms if you don't already know them. You should also go back to your textbook as you read through the chapter for more information on any topics that are frighteningly unfamiliar to you.

In this chapter, we'll start by discussing some important characteristics of populations and then lead you through a short section on how and why populations grow. Next we'll get to the heart of the topic, in a section specifically devoted to human population growth. Remember to use the techniques you learned in Part III as you complete the drills—the more practice you have using those techniques, the better prepared you'll be on test day! Let's begin.

SOME IMPORTANT TERMS USED TO DESCRIBE POPULATIONS

A **population** is defined as a group of organisms of the same species that inhabits a defined geographic area at the same time. Individuals in a population generally breed with one another, rely on the same resources to live, and are influenced by the same factors in their environment.

Two important characteristics of populations are the density of the population and how the population is dispersed. **Population density** refers to the number of individuals of a population that inhabit a certain unit of land or water area. An example of population density would be the number of squirrels that inhabit a particular forest. **Population dispersion** is a little more complicated; this term refers to how individuals of a population are spaced within a region. There are three main ways in which populations of species can be dispersed, and you should know all of them for the test! We've listed them on the following page.

- **Random:** The position of each individual is not determined or influenced by the other members of the population. An example is seen in species of plants that are interspersed in fields or forests—the location of their growth is random, and relative to other species, not their population. This type of dispersion is relatively uncommon.

- **Clumping:** The most common dispersion pattern for populations. In this type of dispersion, individuals "flock together." This makes sense for many species—certain plants will all grow together in a region that suits their requirements for life; fish swim in schools to avoid predation; and birds and many other animals migrate in groups.

- **Uniform:** The members of the population are uniformly spaced throughout their geographic region. This is seen in forests, in which trees are uniformly distributed so that each receives adequate light and water, for example. Uniform dispersion is often the result of competition for resources in an ecosystem.

Got it? Now let's look at how populations grow (and shrink!).

POPULATION GROWTH

So, we know what populations are and how they're dispersed, but how do populations grow? What determines if they will or will not grow? When they grow, *how* do they grow? These are all questions that you'll need to be able to answer on test day. Let's review the basics of population size and growth before we get into a more specific discussion of how human population growth occurs.

The **biotic potential** of a population is the amount that the population would grow if there were unlimited resources in its environment. This is not a practical model for population growth simply because in reality the amount of resources in the environments of populations are limited.

As we reviewed in the last chapter, in every ecosystem, members of a population compete for space, light, air, water, and food. The **carrying capacity** (K) of a particular region is defined as the maximum population size that can sustainably be supported by the available resources in the region. As you might expect, geographic regions have different carrying capacities for populations of different species—because different species have different requirements for life. For example, in a given area, you would expect a population of bacteria to be quite a bit larger—in terms of the number of individuals—than a population of zebras. This is because individual bacteria are much smaller than individual zebras; thus, each bacterium requires fewer resources to live than each zebra.

Population Growth Graphs

If we looked at the growth of a population of bacteria in a Petri dish with plenty of food, the curve produced by plotting the increase in their number over time would be in the shape of a J, because the bacteria would grow exponentially. The exponential growth curve is shown below.

Exponential (unrestricted) growth

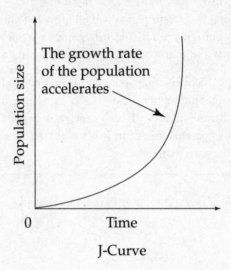

J-Curve

Now, we said that this exponential growth rate is seen in populations in which resources are unlimited, but in nature these ideal settings are rare and fleeting. In a more realistic model for population growth, after the initial burst in population, the growth rate generally drops, and the curve ultimately resembles a flattened S.

This type of growth, which is a much better model for what exists in natural settings, is called **logistic population growth**. The logistic growth model basically says that when populations are well below the size dictated by the carrying capacity of the region they live in, they will grow exponentially, but as they approach the carrying capacity, their growth rate will decrease and the size of the population will eventually become stable. This logistic growth is shown in the graph on the following page.

Logistic (restricted) growth

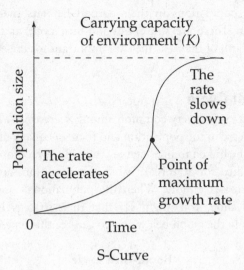

S-Curve

We can predict long-term population growth rates using a model called the Rule of 70. The **Rule of 70** says that the time it takes for a population to double can be approximated by dividing 70 by the current growth rate of the population. For example, if the growth rate of a population is 5 percent, then the population will double in 14 years ($\frac{70}{5}$ percent = 14 years). The Rule of 70 can be used to estimate the number of years for any variable to double, the doubling time.

Not surprisingly, the rate of growth of a population depends on the species that makes up the population. Species can be divided into two groups based on their reproductive strategies: the *r*-selected pattern or the *K*-selected pattern. **r-selected organisms** reproduce early in life and often, and have a high capacity for reproductive growth. Some examples of *r*-selected species are bacteria, algae, and protozoa. In these species, little or no care is given to the offspring, but due to the sheer numbers of offspring in the population, enough of the offspring will survive to enable the population to continue. On the other hand, **K-selected organisms** reproduce later in life, produce fewer offspring, and devote significant time and energy to the nurturing of their offspring. For these species, it is important to preserve as many members of the offspring as possible because they produce so few; parents have a tremendous investment in each individual offspring. Some examples of *K*-selected species are humans, lions, and cows. Many species lie on the continuum between these two strategies, but the groups are useful for broad comparisons.

Population Cycles

When we observe populations in their natural habitats there are two distinct patterns that occur. These are the boom-and-bust cycle and the predator-prey cycle. Lets look at both of them, as they are important for the exam.

Boom-and-Bust Cycle

The boom-and-bust cycle is very common among *r*-strategists. In this type of cycle there is a rapid increase in the population and then an equally rapid drop off. These rapid changes may be linked to predictable cycles in the environment (temperature or nutrient availability, for example). When the conditions are good for growth, the population increases rapidly. When the population's conditions worsen, its numbers rapidly decline. You might say that their strategy is "get it while the getting's good." Study the graph below so you can see this type of cycle in action.

Boom-Bust Cycle

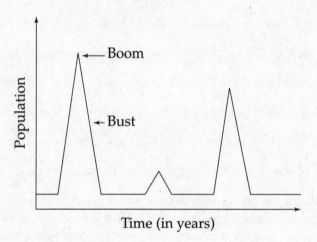

Predator-Prey Cycle

Remember the rabbit and coyote populations from the last chapter? We discussed how in a year of relatively high rainfall, rabbits have plenty of food, which enables them to reproduce very successfully. In turn, because the coyote is a predator of the rabbit, coyotes would also have plenty of food, and their populations would also rise rapidly. However, if the rainfall is below average a few years later, then there would be less grass; the population of rabbits would decline, and the coyote population would decline in turn. The graph of the predator-prey relationship looks like this:

Predator-Prey cycle

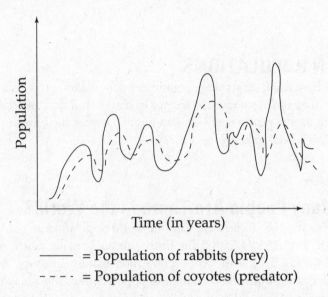

——— = Population of rabbits (prey)

- - - - = Population of coyotes (predator)

Something important to notice in this graph: the population of the coyote does not change at exactly the same time as the rabbit population. The coyote population actually rises *after* the rabbit population does. That is because the rabbit population has to have time to build up to fairly high levels before the coyotes can find enough to eat. When there is enough food, the coyote mothers have enough energy to give birth to and feed their pups. Only then can the coyote population increase.

The predator-prey cycle also plays a role in understanding why many endangered species are large carnivores. Large predator populations can suffer directly if humans alter their natural habitats, but they can also suffer indirectly if humans kill off their prey. If the prey population falls so low that the predator cannot find food, then the predator population will decline, sometimes to the point of extinction.

Factors Influencing Population Growth

There are population-limiting factors that are purely the result of the size of the population itself. For example, in many populations of species in nature, birth and death rates are influenced by the density of the population. Other **density-dependent** factors that influence population size are increased predation (which occurs because there are more members of the population to attract predators); competition for food or living space; disease (which can spread more rapidly in overcrowded populations); and the buildup of toxic materials.

Some population-limiting factors operate independently of the population size. These **density-independent** factors will change the population's size regardless of whether the population is large or small. Independent factors include fire, storms, earthquakes, and other catastrophic events.

Now that you have a basic understanding of how and why populations change in size, let's move on to discuss human populations more specifically.

HUMAN POPULATIONS

You might have heard something about human population growth as you read the news or studied biology and earth science in school. But do you know how many humans are on the planet now? Do you know how fast the human population is growing? You'll need to know for test day.

How Many People Are There in the World?

According to the U.S. Census Bureau, the world population as of March 2014 is estimated to be 7,152,717,500. The birth rate has actually fallen in the United States and worldwide, but the world population is still increasing. Take a look at the following table, which shows estimated populations of some major countries as of the March 2014.

Top 10 Most Populous Countries	
Country	Estimated Population (March 2014)
China	1,355,692,576
India	1,236,344,631
United States	318,892,103
Indonesia	253,609,643
Brazil	202,656,788
Pakistan	196,174,380
Nigeria	177,155,754
Bangladesh	166,280,712
Russia	142,470,272
Japan	127,103,388

Source: U.S. Census Bureau, International Data Base

The following graph shows how the overall world population of humans has increased since 1950, and predicts it will continue to increase into the 2050s.

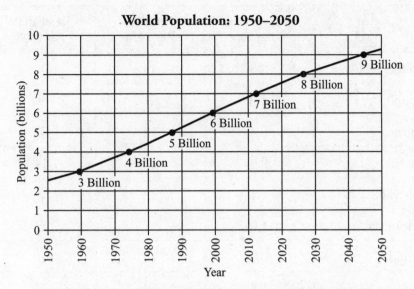

Source: U.S. Census Bureau, International Data Base, December 2010 Update.

The population of many of the countries shown in the table is currently increasing in size. We can determine the rate of population change of a country by using a simple formula, if we only consider the contributions of births and deaths to changes in population size.

$$\text{Actual Growth Rate (\%)} = \frac{\text{birth rate} - \text{death rate}}{10}$$

The birth rate (the **crude birth rate**) is equal to the number of live births per 1,000 members of the population in a year, and the death rate (or **crude death rate**) is equal to the number of deaths per 1,000 members of the population in a year.

The table below shows the 2014 growth rates for 11 countries. Note that some countries are experiencing negative growth.

Country	Growth Rate % Per Year	Birth Rate Per 1,000	Death Rate Per 1,000
United States	0.8	13	8
Japan	-0.1	8	9
China	0.4	12	7
Russia	0.0	12	14
Australia	1.1	12	7
Indonesia	1.0	17	6
Canada	0.8	10	8
Mexico	1.2	19	5
South Africa	-0.5	19	17
India	1.3	20	7
United Kingdom	0.5	12	9

Source: U.S. Census Bureau, International Data Base

How Do Human Populations Change?

Populations can also change in number as a result of migration into and out of the population. Two important vocabulary words to describe human migration are **emigration,** which is the movement of people out of a population, and **immigration,** which is the movement of people into a population. Keep in mind that, in general, emigration and immigration are only small factors in the changes in size of human populations.

The most significant additions to human populations are due to births, plain and simple. The term **total fertility rate** is used to describe the number of children a woman will bear during her lifetime, and this information is based on an analysis of data from preceding years in the population in question. Total fertility rates are predictions that provide a rough estimate, but they can't be depended on because they assume that the conditions of the past will be the conditions of the future.

The **replacement birth rate** of a human population refers to the number of children a couple must have in order to replace themselves in a population. While you might automatically think that the answer is always two, in reality it is slightly higher to compensate for the deaths of children, the existence of non-child bearing females in the population, and other factors. In developing countries, the replacement birth rate can be as high as 2.5!

As we mentioned earlier, despite the relatively recent drop in total fertility rates worldwide, the world's population is still increasing. This is because many members of the human population are future parents, so even if they only reproduce at a replacement rate, there will be an overall increase in the total population. Now, let's look at a graph of how the overall world population growth rate has changed and is projected to change from 1950–2050.

World Projected Population Growth Rate: 1950–2050

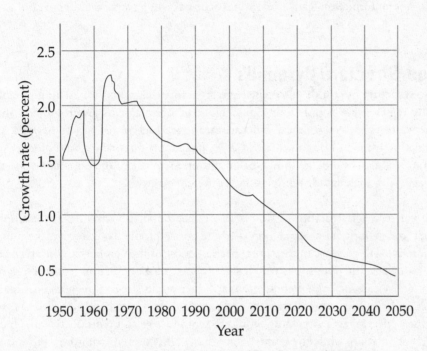

Source: U.S. Census Bureau, International Data Base, April 2005 version.

Though not shown on this graph, the human population has actually been growing exponentially for more than three centuries. But what did we learn earlier in this chapter? That no population can grow exponentially indefinitely... we'll talk more about that in a bit. For now, let's discuss some factors that affect the growth rates of human populations.

As you can see, perhaps not surprisingly, there is a strong empirical correlation between the education level of women and the growth rate of populations. Additionally, the reason that religion and culture are predictors of birth rates is that in some countries, certain groups have a proclivity toward reproduction for religious reasons.

The reason that the world's population has grown so considerably, especially in the past 100 years, is not because of an increased number of births, but because of the significant drop in the world death rate. People are living longer lives, and there are far

Not surprisingly, a number of factors affect the total fertility rates in a population, and as a result, the population's birth rate. Among these are

- The availability of birth control

- The demand for children in the labor force

- The base level of education for women

- The existence of public and/or private retirement systems

- The population's religious beliefs, culture, and traditions fewer infant deaths today than there were 100 years ago. This is due, in large part, to the Industrial Revolution, which improved the standard of living for millions living in industrialized nations. Other causes of the extension of the human life span are the development of clean water sources and better sanitation, the creation of dependable food supplies, and better health care. In general, the overall health of a population can be estimated by examining the expected life span of individuals and the mortality rate of infants.

Life-span information leads us right into our next topic: age-structure pyramids.

Age-Structure Pyramids

Age-structure pyramids (also called **age-structure diagrams**) are useful for graphically representing populations. Some age-structure diagrams group humans into three categories by age: those who are **pre-reproductive** (0–14), those who are **reproductive** (15–44), and those who are **post-reproductive** (45 and older). Age pyramids, such as the one shown below, group members of the population strictly by age, with each decade representing a different group.

We can use age-structure pyramids to predict population trends; for example, when the majority of a population is in the post-reproductive category, the population size will decrease in the future because most of its members are incapable of reproducing. The opposite is true if the majority of a population is in the pre-reproductive category; these populations will increase in size as time goes on. Take a look at the age-structure diagram for Mexico and the United States, below. As you can see, Mexico has a large number of pre-reproductive and reproductive members in its population, while the United States has a fairly even distribution. From this, we can see that the population of Mexico should increase significantly over time, while the population of the United States should grow more slowly.

Age-Structure Pyramid

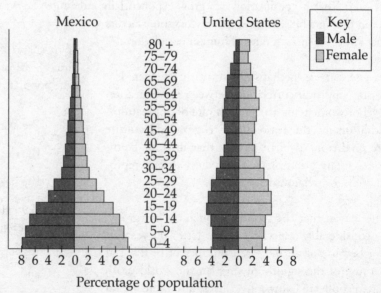

The Demographic Transition Model

The **demographic transition model** is used to predict population trends based on the birth and death rates of a population. In this model, a population can experience zero population growth via two different means: as a result of high birth rates and high death rates; or as a result of low birth rates and low death rates. When a population moves from the first state to the second state, the process is called **demographic transition.** The four states that exist during this transition are:

1. The **pre-industrial state:** In this state, the population exhibits a slow rate of growth and has a high birth rate and high death rate because of harsh living conditions. Harsh living conditions can be considered **environmental resistance**, an umbrella term for conditions that slow a population's growth.

2. The **transitional state:** In this second state, birth rates are high, but due to better food, water, and health care, death rates are lower. This allows for rapid population growth.

3. The **industrial state:** In the third state, population growth is still fairly high, but the birth rate drops, becoming similar to the death rate. Many developing countries are currently in the industrial state.

4. The **postindustrial state:** In the final state, the population approaches and reaches a zero growth rate. Populations may also drop below the zero growth rate (as we saw in Russia, South Africa, and Japan, in the table of growth rates).

Check out a graph to better understand the demographic transition model.

Demographic Transition Model

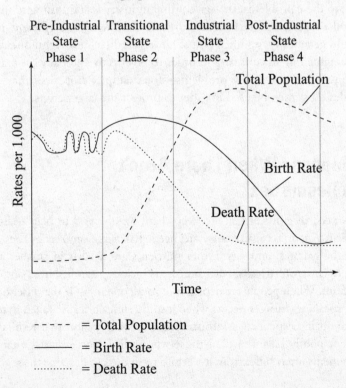

You will most likely see questions that will ask you specifically to describe the demographic transition model, or apply it to hypothetical or real population states, so make sure you understand this material—and if you don't, go back to your textbook for further explanation!

THE IMPACT OF HUMANS ON EARTH

As you're probably well aware, humans have the greatest impact on the environment of any living species on Earth, and the increase in our population over the last few centuries has seriously and dramatically changed the face of the earth.

Four of the most significant factors that have contributed to the increase in the world human population are: improved nutrition, the availability of clean water, newly implemented systems for sanitary waste disposal, and better medical care.

Another very significant factor has been the increase in food production. Almost half of Earth's land surface is currently devoted to various ways of producing food for humans; specifically, 12 percent of the earth's land is composed of farms, 11 percent is composed of forests planted by humans, and 26 percent is used for grazing livestock! As we mentioned earlier, this enormous amount of food production takes its toll on the land, and we now face excessive and harmful erosion, in addition to a variety of environmental problems that have resulted due to the wide-scale use of irrigation. Finally, the widespread use of pesticides and fertilizers for increased crop yields leaves large amounts of harmful chemical residues in the soil and water.

In response to these problems, the agricultural industry is continuing to invent and promote soil conservation techniques, organic farming, more efficient irrigation methods, and genetically modified crops. However, these new techniques will need to be implemented in all countries in order to be effective globally. In many cases, new techniques introduce new problems—for example, crop sizes may increase, but pesticides must then be used in order to protect the larger crops.

What Happens When There Aren't Enough Resources?

Our bodies need certain nutrients to keep them healthy and to help resist disease. Some nutrients, or **macronutrients,** are needed in large amounts. These include proteins, carbohydrates, and fats. Other nutrients are needed in smaller amounts; these are called **micronutrients**. Micronutrients include vitamins, iron, and minerals such as calcium. When people are deprived of food, one result is the onset of hunger. Technically speaking, **hunger** occurs when insufficient calories are taken in to replace those that are being expended. **Malnutrition** is poor nutrition that results from an insufficient or poorly balanced diet; those whose diets lack essential vitamins and other components often suffer from it. A third term used to describe those who aren't receiving sufficient resources is **undernourished.** Undernourished people have not been provided with sufficient quantity or quality of nourishment to sustain proper health and growth.

According to the Food and Agriculture Organization, or FAO, 840 million people on Earth are hungry. Some 799 million of these people are living in developing countries, but, perhaps surprisingly, the remaining 41 million are living in developed nations. On the other hand, 30 percent of the total population of the United States is considered obese (which means that they are more than 30 lbs. overweight) and globally, 1.2 billion people are overweight. Why does this dichotomy exist?

While the reasons for hunger are many and complex, the simplest answer to this question is: poverty. Our planet produces sufficient food to feed today's world population, but many people lack the money to buy food or the resources to produce it.

All over the world, human communities are trapped in a cycle of poverty, resource degradation, and high fertility. For example, in the first third of the twentieth century, Asia, Africa, and Latin America produced enough grain that it was not necessary for them to import it from other countries. However, because of their constantly increasing populations, all of these countries are now importing grain; this is an ominous sign of impending problems with hunger in these countries.

Encouragingly, China, Thailand, and Indonesia are working hard to implement government reforms that will increase the quality of life for their citizens. In China, for example, as a result of reform and development in rural areas, the number of people in the country without enough food and clothing has decreased from 250 million in 1978, to 29 million in 2004.

Mass Extinction

During the past 500 million years, there have been five major extinction events, during which a significant percentage of all life on Earth including the oceans was exterminated by some catastrophic event. The most famous incident was the asteroid impact 65 million years ago that eliminated the dinosaurs and other large reptiles, and vacated numerous niches that were quickly filled by mammals through radiative adaptation. The lesson to gain from this information is that life on Earth is very adaptive and capable of surviving great stress. Presently, humans are stressing the planet and may render Earth uninhabitable for our species. Life, however, will certainly continue.

Hunger in America

Despite its high obesity rate, there are hungry people even in the United States, one of the richest countries in the world—lots of hungry people. In fact, the number of hungry people in the United States is greater now than it was when international leaders set hunger-cutting goals at the 1996 World Food Summit. At this summit, government leaders pledged to cut the number of Americans living in hunger from 30.4 million to 15.2 million by 2010, but this goal has not nearly been met. Today more than 35 million Americans are considered "food insecure." Additionally, at least four in ten people in the United States who are between the ages of 20 and 65 rely on food stamps. That's nearly half of all citizens of the United States!

But why are American people hungry when they live in a nation that's known as the world's breadbasket? Again, the main reason is poverty.

In the mid-1990s, a call for welfare reform resulted in the passing of the Personal Responsibility and Work Opportunity Reconciliation Act (PRWORA). The premise of this welfare reform, according to its proponents, was that people who are able to work should be encouraged to find employment, so that they will not remain dependent on government assistance. The act limited the number of people who qualified for food stamps—and also limited the duration that people could receive food stamps and public support.

While at the time it was generally agreed that welfare reform was necessary, many families are now reaching their deadline for public assistance. Since the implementation of this act, and in the future, it will be of crucial importance for state and local groups to find ways to support the truly needy.

In the United States, there are a number of charitable agencies that provide food at no or low cost to those in need. One example is **Second Harvest,** which makes use of food that would otherwise go to waste. Second Harvest receives food from foods processors and distributors and redistributes it via food banks. Recent federal legislation called the Charity, Aid, Recovery, Empowerment Act of 2003 (the CARE Act), has supported Second Harvest and other food distribution charities by allowing farmers, ranchers, and restaurant owners to deduct the cost of the foods they donate.

Global Hunger Policies—For Good or Ill

While social reform is a viable solution to the problem of hunger in the United States, often the only solution in developing countries is to enable communities to become self-sufficient in food procurement. These destitute communities either need monetary resources that will allow them to purchase necessary food supplies, or the resources that would enable them to produce their own food.

Another issue that must be dealt with in the struggle against global hunger is that many countries just can't produce the food they need in order to feed their citizens. In these cases, the first viable solution mentioned above—that of providing monetary resources to people so they can purchase food—is rendered irrelevant.

The World Trade Organization (WTO) controls the policies of international trade. Unfortunately for smaller, poorer nations, the economically strong nations of the world have more influence over the creation of policies by the WTO. Oftentimes, trade policies undercut prices for developing nations, which makes it difficult for them to enter the world market.

Another problem for developing countries is that a trade imbalance exists between them and developed nations. In many poorer nations, national resources have been degraded in an effort to reduce the national debt. The people in these poverty-stricken nations have nothing to export—except labor. Relatively recently, companies in the United States and other developed nations have begun outsourcing jobs to developing countries. Many Americans are appalled at the terrible conditions under which these overseas laborers work—their hours are abnormally long and they are paid almost nothing. Despite these conditions, the competition between poor communities to secure contracts with companies overseas is fierce because often the alternative is continued poverty.

So, Where Do All These People Live?

The majority of humans live in some type of community, and the largest percentage of the human population lives in relatively large communities and urban centers.

Since the development of ancient civilizations, humans have lived together in large centralized communities, or cities. A couple of ancient cities that you may be familiar with are Rome (in Italy) and Athens (in Greece). However, never before have the urban centers of the world grown as quickly as they are growing now. If we traced the growth of urban areas in the United States, we would find that before the Civil War (around the 1850s), only about 15 percent of the population lived in a city. Around the time of World War I (1920), that number grew to encompass about 50 percent of the total population of the United States, and today it hovers around 75 percent.

Globally, almost half of the world's population today lives in an urban area. In the United States, this is partly due to the fact that our aging population has largely moved into the cities to have greater access to health services, employment opportunities, and cultural activities.

When considering those who live in urban areas, we also count those who reside in satellite communities, or **suburbs.** In recent times, with lower oil prices making it easier for people to afford gasoline to commute to and from their jobs in cars, people have moved out of city centers in order to have more living space. Interestingly, people who live in the suburbs, on average, occupy eleven times more space than do those who live in the city. One of the advantages of living in the suburbs is that people have their own land space—a backyard—which they need not share with others.

The term used to describe the emigration of people out of the city and into the suburbs is **urban sprawl.** In some areas of the United States, urban sprawl takes over vast tracts of land. In Colorado, for example, population growth has resulted in a number of new communities between Denver and Boulder; when traveling between the two cities, it is now difficult to determine where the Denver metro area ends and the city of Boulder begins.

When urban areas grow too large and become too dense, providing water to all of their citizens becomes increasingly difficult. Coupled with this is the strain on the water supply—more people means more water use. In many of these newly crowded areas, water shortages have led to the implementation of restrictions on water usage.

Another problem that results from the increase in the populations of cities is what to do with all of the waste that's created. When you think about it, almost all human activities create waste—when you go to your local coffee shop and get a cup of coffee, you probably don't think much of it. However, if you get a cup of coffee every morning in a paper cup, five days a week for the 52 weeks of the year, then at the end of the year you've accumulated more than 250 cups! That's a significant pile of garbage.

Transportation Alternatives

While many people find the suburbs a pleasant place to live, ecologists and city planners have recently come to realize that this urban sprawl may reduce quality of life for all urban dwellers. One major concern of policy makers and citizens in metro areas is what to do about transportation. Ideally, people would be encouraged to use mass transit or participate in car pools rather than drive separately in private cars. Having fewer cars on the road decreases the amount of air pollution from automobile emissions and also results in less congestion on roadways.

Other environmentally conscious or "green" modes of transportation include bicycles, motor scooters, and electric bikes. Larger cities often opt to build subway systems, but they are extremely expensive to develop and are only cost-effective when there are enough people who will pay to use them. However, city buses are an option for both large and small cities; fleets of buses are less expensive than subways to create and maintain, and although they contribute to road congestion, they decrease congestion by having more people per vehicle.

Rapid-rail or light rail systems are more common in Japan and Western Europe than the United States, but as of recently they're being considered as an option for cities that lack subways. Rapid rail systems work by magnetic levitation; suspended above a track, a train moves along as a result of strong attractive and repulsive magnetic forces.

Building Sustainable Cities

In order for cities to be sustainable, city planners and developers must build and manage cities to work with, and within, their natural settings, instead of merely placing buildings and structures in these settings.

There are certain cities in the United States and elsewhere in the world that are setting examples of progressive thinking in conservation and ecology. For example, the city of Boulder, Colorado has long been recognized for its forward-thinking, green policies. Bicycle paths cover the city, allowing cyclists to move freely from one area of the community to another. Buses move around the city and in and out of Denver and the surrounding communities, which enables people to commute to work without using their cars and creating more emissions. For those who need to drive to work, the city encourages carpools and provides parking areas for those who carpool. Additionally, the city's strong recycling programs help reduce the amount of material that's added to land fills. The city is ringed by open spaces that can be used by the city population for recreation. These areas are leased to local ranchers for grazing cattle. Boulder's citizens have voted to tax themselves to pay for the purchase of additional communal green space for the city!

Other cities that have been held up as models for city planning are Curitiba, Brazil and Portland, Oregon. Curitiba has an excellent mass transport system, as well as bicycle paths and pedestrian walk ways. The city provides recycling programs, job training, health care, and environmental education for its citizens. Likewise, in the 1970s the state of Oregon became determined to head off urban sprawl, and the city of Portland established statewide zoning policies and restrictive growth policies for urban areas. City developers were encouraged to invest in established neighborhoods rather than develop undisturbed areas. The city of Portland established Metro, a regional body that deals with land use, city planning, and the development of natural areas. Light rail systems were developed, and Metro began to encourage neighborhood self-sufficiency in order to keep the number of people who need to commute for food or other supplies to a minimum.

Now and in the future, it will be important for city planners to deal with new problems created as a result of urban sprawl. City planners and developers must take environmental concerns into consideration; providing green spaces and transportation alternatives, and planning for the supply of water are all relatively new challenges for those involved in building cities.

Big Cities in Less Developed Countries

So far we've made it sound like the cities of the world are dealing quite well with the boom in their population, but this is not true globally. In less developed countries, the increase in the population size of major cities has many very negative effects. Among the worst effects is a deficiency of housing or habitable areas for the burgeoning population. As a result, people are homeless, become "squatters," or make their homes in areas that are completely undeveloped—areas that have no water or electricity, or stable, durable housing.

Some of the reasons people in less developed countries are moving to cities are similar to those of people in developed countries; for example, cities have more opportunities for employment. However, these people often have other motivations that drive them out of the country, such as war, religious or cultural persecution, or the degradation of their environment.

Ecological Footprint

One concept that you should definitely be familiar with for this exam is that of the ecological footprint. An **ecological footprint** is used to describe the environmental impact of a population. It is defined as the amount of the earth's surface that's necessary to supply the needs of, and dispose of the waste of, a particular population. Americans have one of the largest ecological footprints; we require about 9.7 hectares per capita (per person). One hectare is 10,000 square meters, or about 2.5 acres. America's amount is comparatively enormous—the ecological footprint of Indonesia is only 1.1 hectares per capita. In general, affluent populations have a much higher ecological footprint than non-affluent ones.

We can use a mathematical model to describe the impact that humans have on the environment. Nicknamed the IPAT model, it is written as

$$I = P \times A \times T$$

In the model, I = the total impact, P = population size, A = affluence, and T = level of technology.

While you probably won't be asked to calculate the impact of populations on this exam, it's a good idea to know this formula exists, and that the variables of population size, affluence, and level of technology all influence the impact a population has on its environment.

Threatened and Endangered Species

Another way humans impact Earth is through their interaction with animals. Human activities have caused or contributed to the **extinction** of many species. Some species are **threatened**, meaning the number of individuals of a species is quite low, or they are **endangered**, meaning the species is in imminent danger of going extinct. In 2004, the World Conservation Union estimated that there were approximately 15,500 worldwide species that were endangered, ranging from 12 percent of the bird species to 42 percent of the tortoises. Plants and organisms living in marine and freshwater ecosystems face similar problems. The United Nations estimates that one in five species of coral reefs (or 200,000 species) are in danger of becoming extinct in the next 40 years. Extinctions have happened throughout Earth's history—this natural rate of extinction is called the **background extinction rate**. Knowledgeable scientists estimate that the current extinction rate is between 50 and 500 times higher than in the past, probably due to human influence. Extinctions can happen anywhere in the world, but the rates are particularly high in the tropics (mostly mountains and islands).

The species that are most endangered have several factors in common: they require large ranges of habitat to survive, have low reproductive rates, have specialized feeding habits, and have low population numbers.

Humans play a major role in the extinction of species because of our destruction of animal and plant **habitats**. Poverty and rapid population growth cause people to use destructive practices, such as slash and burn farming, that destroy species' habitats. When we build roads or cities, habitats are lost or **fragmented** (broken into smaller pieces). Finally, we cause habitat **degradation** by adding pollutants to the environment. Other factors that can contribute to extinction are invasive species and direct hunting or **overexploitation** for animal products. Dr. Norman Myers coined the term **biodiversity hot spot** to describe a highly diverse region that faces severe threats and has already lost 70 percent of its original vegetation.

There are things we can do to reduce the rate of extinction. Living sustainably and conserving resources helps lower the demand that destroys habitats. Making it illegal to trade in specific organisms means that those organisms will not be hunted or collected. We can also help organisms on a species-by-species approach. Zoos and other institutions have captive breeding programs in which endangered species are bred under human control until their populations are high enough to be reintroduced into the wild. We can conserve habitats by requiring that large tracts of land be set aside and protected from human activity. In these protected habitats, organisms will find their niche and survive without risk of human interference. National parks and animal sanctuaries are two examples of protected habitats.

There are many United States laws that have been passed to reduce the rates of extinctions and protect specific organisms. Three very important ones are:

Date	Law	What It Did
1972	Marine Mammal Protection Act	This act protected marine mammals from falling below their optimum sustainable population levels.
1973	Endangered Species Act Program for the protection of threatened plants and animals and their habitats	The act prohibited the commerce of those species considered to be endangered or threatened.
1973	Convention on International Trade in Endangered Species of Wild Flora and Fauna (CITES)	This agreement bans the capture, exportation, or sale of endangered and threatened species.

Use the acronym **HIPPCO** to memorize the causes of extinction:

- habitat destruction/fragmentation
- invasives
- population
- pollution
- climate change
- overharvesting

Now you're ready for the key terms review and the following drills. Remember to use our techniques as you go through them!

KEY TERMS

Know these terms backward and forward so that you can spout them in your sleep.

Population Growth

population density
population dispersion: random, clumping, uniform
biotic potential
carrying capacity
logistic population growth
Rule of 70
r-selected, K-selected
boom-and-bust cycle
predator-prey cycle
density-dependent and independent factors

Human Populations

crude birth rate
crude death rate
emigration
immigration
total fertility rate
replacement birth rate
age-structure pyramids
pre-reproductive, reproductive, post-reproductive
demographic transition: pre-industrial, transitional
        state, industrial state, post-industrial state
environmental resistance
malnutrition
macronutrient
micronutrient
undernourished
second harvest
suburbs
urban sprawl
ecological footprint
IPAT model

Threatened and Endangered Species

extinction
threatened
endangered
background extinction rate
fragmented habitat
habitat degradation
overexploitation
biodiversity hot spot
CITES
HIPPCO

Chapter 6 Drill

Directions: Each of the questions or incomplete statements below is followed by five suggested answers or completions. Select the one that is best in each case. For answers and explanations, see Chapter 13.

1. Populations have all the following characteristics EXCEPT

   (A) density
   (B) dispersion
   (C) habitat
   (D) gene pool
   (E) size

2. Which of the following describes individuals leaving a population?

   (A) Birth rate
   (B) Carrying capacity
   (C) Immigration
   (D) Emigration
   (E) Environmental resistance

3. A population has a growth rate of 2 percent per year. How long will it take for this population to double?

   (A) 70 years
   (B) 40 years
   (C) 35 years
   (D) 15 years
   (E) 2 years

4. An age-structure pyramid is used to

   (A) study the immigration rates in a population
   (B) calculate the doubling time of a population
   (C) study the carrying capacity of a habitat
   (D) determine what the density-dependent factors are for a population
   (E) study the number and ages of people in a country

5. Which of the following are exhibited by k-select organisms?

   I. Slow maturation
   II. Many small offspring
   III. Reproduction occurs late in life
   (A) I only
   (B) II only
   (C) III only
   (D) I and II only
   (E) I and III only

6. A population cycle that is marked by regular increases and decreases in its numbers is correctly said to be

   (A) boom-and-bust
   (B) irruptive
   (C) stable
   (D) logistic
   (E) irregular

7. The demographic transition model is used to study the

   (A) effects of migration patterns
   (B) influence of industrialization on population growth or decline
   (C) location of large population centers
   (D) benefits of mass transportation projects
   (E) negative effects of pollution on the habitat

8. Which disease is having a severe negative impact on the population in sub-Saharan Africa today?

   (A) Lung cancer
   (B) Heart disease
   (C) HIV/AIDS
   (D) Alzheimer's
   (E) Down syndrome

9. Which of the numbers below is closest to the population of India?

   (A) 1 billion
   (B) 900 million
   (C) 300 million
   (D) 50 million
   (E) 2 million

10. Which of the following is a density independent population factor?

    (A) Number of parasites in the population
    (B) Number of predators in the population
    (C) Competition for resources
    (D) Disease
    (E) Habitat destruction

11. When a population encounters environmental resistance it is most likely to

   (A) continue its high growth rate
   (B) mutate to form and continue growing
   (C) slow down its growth rate
   (D) move to a higher growth rate
   (E) have no effect on the growth rate

12. A population's growth can best be calculated using which of the following?

   (A) Births + immigration – deaths + emigration
   (B) Immigration + emigration
   (C) Emigration + births
   (D) Births + emigration – deaths + immigration
   (C) Immigration – emigration

13. Overexploitation of a species can happen by all of the following Except

   (A) excessive hunting
   (B) use of a species for food
   (C) use of species as a pet
   (D) habitat destruction
   (E) habitat conservation

Directions: Each set of lettered choices below refers to the numbered questions or statements immediately following it. Select the one lettered choice that best answers each questions or best fits each statement. A choice may be used once, more than once, or not at all in each set.

Questions 14-18 refer to the following characteristics of populations.
   (A) birth rate
   (B) total fertility rate
   (C) mortality rate
   (D) life expectancy
   (E) replacement birth rate

14. The number of people who die per 1,000 in the population

15. The average number of years a person can be expected to live

16. The average number of offspring a woman is expected to have

17. The number of individuals born per 1,000 in the population

18. The number of children a couple must have to replace themselves

19. Poverty can affect population in all of the following ways EXCEPT

   (A) causing premature deaths
   (B) increasing total fertility rate
   (C) decreasing total fertility rate
   (D) forcing the use of resources in unsustainable ways
   (E) emigration

Free-Response Question

1. A habitat's carrying capacity imposes limits on the growth of populations and their consumption of resources.

   (a) Define the term "carrying capacity." Give two examples of how carrying capacity can impose limits on a population.

   (b) Explain how a population's consumption of natural resources might be controlled. Give two examples of how nature slows down the consumption of natural resources by a population.

   (c) Describe two ways human activity can raise a habitat's carrying capacity for humans.

REFLECT ACTIVITY

Respond to the following questions:

- For which content topics discussed in this chapter do you feel you have achieved sufficient mastery to answer multiple-choice questions correctly?

- For which content topics discussed in this chapter do you feel you have achieved sufficient mastery to identify hot-button topics and successfully articulate them in free-response questions?

- For which content topics discussed in this chapter do you feel you need more work before you can answer multiple-choice questions correctly?

- For which content topics discussed in this chapter do you feel you need more work before you can successfully identify and articulate them in free-response questions?

- What parts of this chapter are you going to re-review?

- Will you seek further help, outside of this book (such as a teacher, tutor, or AP Central), on any of the content in this chapter—and, if so, on what content?

Chapter 7
Resource
Utilization

According to the AP topic outline for environmental science, about 10–15 percent of the questions you'll see on the exam will ask about subjects we'll review in this chapter. Let's take a step back and review the fundamental themes of this chapter. First of all, a "**resource**" is strictly defined as "an available supply that can be drawn on as needed." That's easy enough to understand, but what resources are we talking about? Well, all of the earth's resources that humans rely on to live—namely the land, water, and the things that grow from them. As we'll see in this chapter, humans use the land and water for countless reasons.

We will begin our discussion with a description of the resources of the world—including what happens if people don't get enough resources, and who has too few. We'll then go through the resources gleaned from agriculture, forests, oceans, and mining. We'll end with a (short!) discussion of the economics behind our resource use. Let's begin!

SHARE AND SHARE ALIKE?

When people talk about managing common property resources such as air, water, and land, a paper published in *Science* magazine by Garret Hardin in 1968, called "The Tragedy of the Commons," often comes to mind. In this paper, Hardin referenced a parable from the 1880s in which a piece of open land, a commons, was to be used collectively by the townspeople for grazing their cattle. Each townsperson who used the land continued to add one cow or ox at a time until the common was overgrazed. Hardin quite eloquently says, "Each man is locked into a system that compels him to increase his herd without limit—in a world that is limited. Ruin is the destination toward which all men rush, each pursuing his own best interest in a society that believes in the freedom of the commons. Freedom in a commons brings ruin to all."

This parable serves as a foundation for modern conservation. **Conservation** is the management or regulation of a resource so that its use does not exceed the capacity of the resource to regenerate itself. This is different from **preservation,** which is the maintenance of a species or ecosystem in order to ensure their perpetuation, with no concern as to their potential monetary value.

In this chapter, we'll continue to show how human economics often influences how we interact with Earth's resources. Ecosystems (both biotic and abiotic) are often referred to as **natural resources**. When we describe something as a *resource*, we are essentially putting an economic value on it; therefore, natural resources are described in terms of their value as **ecosystem capital.**

Some Terms Used to Describe Resources

Let's start by discussing the two main types of resources.

- **Renewable resources** are resources such as plants and animals. These resources can be regenerated quickly. The time necessary for hardwood trees to mature (about 50 years) is widely considered the cross-over point from renewable resources to non-renewable resources. Soil is a special renewable resource. When compared to other renewable resouces, it may seem not that renewable because soil's rate of increase is about 1 inch per 500 years. That is slow, but still renewable.

- **Nonrenewable resources** are things like minerals and fossil fuels. Nonrenewable resources are typically formed by very slow geologic processes, so weconsider them incapable of being regenerated within the realm of human existence.

There are a couple more terms you should know before we dive into our review of the major resources available to humans on Earth; these are consumption and **production.** The **consumption** of natural resources refers to the day-to-day use of environmental resources such as food, clothing, and housing. On the other hand, **production** refers to the use of environmental resources for profit. An example of this might be a fisherman who sells his fish in a market. Got those terms? Let's move on.

AGRICULTURE

How do resources relate to your dinner? Well, 77 percent of the world's food comes from croplands, 16 percent comes from grazing lands, and 7 percent comes from ocean resources. Despite the importance of our crops, and although the population of the United States has increased significantly, fewer people than ever in the history of the United States now farm the land. Why is this? The short answer is that it has a lot to do with increasing urbanization (as we mentioned in the last chapter) and industrialization. Now that machines are readily available to work the land and harvest crops, farms have become more like factories—currently only 15 percent of the entire workforce of the United States produces the food to feed the entire country and for food exporting. Farms in the United States today are quite a bit larger than farms of the past; the average farm is 400 acres, while in the early twentieth century the average farm size was about 100 acres.

The use of machinery in farming has allowed a farmer to work more land more efficiently; however, one of the drawbacks of the machinery is the amount of fossil fuel needed to power it. As the cost of fuel rises, the cost of food will also rise.

This rise in agricultural productivity can be tied to the new pesticides and fertilizers, expanded irrigation, and the development of new high-yield seed types. However, it has also resulted in a significant decrease in the genetic variability of crop plants and led to huge problems in erosion.

Traditional Agriculture and the Green Revolution

Throughout most of history, agriculture all over the world was such that each family grew crops for themselves, and families primarily relied on animal and human labor to plant and harvest crops. This process is called **traditional subsistence agriculture,** and it provides enough food for one family's survival. Traditional subsistence agriculture is still practiced in developing nations, and is currently practiced by about 42 percent of the world's population.

Another component of traditional agriculture that's still practiced in many developing countries today is a method called **slash and burn;** this practice actually dates back to early man. In slash and burn, an area of vegetation is cut down and burned before being planted with crops. Then, because soils in these developing countries are generally poor, the farmer must leave the area after a relatively short time and find another location to clear. This practice severely reduces the amount of available forest; it is a significant contributor to deforestation.

Do not confuse the Green Revolution, which is about farming, with the Green Movement, which is about conservation.

The **Green Revolution** is generally thought of as the time after the Industrial Revolution when farming became mechanized and crop yields in industrialized nations boomed.

Fertilizers and Pesticides

One factor that contributed to the Green Revolution was an increase in the use of fertilizers and pesticides. Interestingly, when the non-native settlers (the first white settlers) planted their first corn crops, certain tribes of Native Americans taught them to plant fish leftovers (the inedible parts of the fish) along with the corn seed; the fish acted as a natural fertilizer for the crops. As you can see, manures and other organic materials have been used as fertilizers by farmers for many years. However, the development of inorganic (chemical) fertilizers brought about the huge increases in farm production seen during the Green Revolution. It's estimated that if chemical fertilizers were suddenly no longer used, then the total output of food in the world would drop about 40 percent!

Of course, there are downsides to the widespread use of chemical fertilizers, and these include the following: the reduction of organic matter and oxygen in soil; the fact that these fertilizers require large amounts of energy to produce, transport, and supply; and the fact that once they are washed into watersheds, they are dangerous pollutants.

Similarly, the increased use of fertilizers and pesticides in the Green Revolution has significantly reduced the number of crops lost to insects, fungi, and other pests, but these chemicals have also had an effect on ecosystems in and surrounding farms. It's estimated that the average insect pesticide will only be useful for 5–10 years before its target pest evolves to become immune to its effects; therefore, new pesticides must constantly be developed. However, even with this constant development, crop loss due to pests has not decreased since 1970, although the used of pesticides has tripled!

Because the use of pesticides is so prevalent in the United States, Congress passed the *Federal Insecticide, Fungicide, and Rodenticide Act (FIFRA)* in 1947 and amended it in 1972. This law requires the EPA to approve the use of all pesticides in the United States. You will need to know this for the exam!

Irrigation

As we've mentioned, another major contributor to the increased crop yields seen in the Green Revolution was advanced irrigation techniques, which allowed crops to be planted in areas that normally would not have enough precipitation to sustain them. However, repeated irrigation can cause serious problems, including a significant buildup of salts on the soil's surface that make the land unusable for crops. To combat this **salinization** of the land, farmers have begun flooding fields with massive amounts of water in order to move the salt deeper into the soil. The drawback to this, however, is that the large amounts of water can waterlog plant roots, which will kill the crops. This process also causes the water table of the region to rise.

Integrated Pest Management (IPM)

When dealing with pests, **integrated pest management** is a more environmentally sensitive approach than chemical pesticides that uses a combination of several methods. Rather than try to get rid of every single pest on the farm, IPM tries to keep the pest population down to an economically viable level. Some of the methods include introducing natural insect predators to the area, intercropping, using mulch to control weeds, diversifying crops, crop rotation, releasing pheromone or hormone interrupters, using traps, and constructing barriers. People using IPM consider using chemical pesticides only in the worst case scenario.

Genetically Engineered Plants.

The third and last significant contributor to the Green Revolution was the introduction of genetically engineered plants. In genetic engineering, scientists try to improve plants by adding genes from one species to another to encourage desirable characteristics, such as longer shelf life, disease/drought/pest resistance, faster growth, and higher crop yields. One beneficial example of this method was the development of golden rice, which contains vitamin A and iron. The introduction of this rice addresses two of the serious health problems that are seen in developing nations: vitamin A deficiency, which can result in blindness and other serious health problems; and iron deficiency, which leads to anemia.

However, there are many problems that arise from **genetically modified organisms** (GMO) as well. Therefore, GMOs have become a very controversial topic. Because this is a relatively new technology, scientists don't know exactly how GMOs will affect the planet ecologically. Genetically modified plants discourage biodiversity, which may harm beneficial insects and organisms, could pose new allergen risks, may increase antibiotic resistance, and could cause new pesticide-resistant pests. Many farmers and consumers are also concerned that cross pollination can contaminate other crops, including organic farms that choose not to use GMOs, or cause unwanted mutations with unknown results.

Monotonous Monoculture

Believe it or not, three grains provide more than half of the total calories that are consumed worldwide! These three crops are rice, wheat, and corn, and the phenomenal increase in the yield of these crops was a result of genetic engineering. Genetic engineers discovered a way to cause plants to divert more of their photosynthetic products (called **photosynthate**) to becoming grain biomass rather than plant body biomass.

It's estimated that, of the roughly 30,000 plant species that could possibly be used for food, only 10,000 have been used historically, with any regularity. Today 90 percent of the caloric intake worldwide is supplied by just fourteen plant species and eight terrestrial animal species! In other words, today's agriculture represents a major reduction in agricultural biodiversity.

On a smaller scale, much of the farming that occurs today is characterized by **monoculture.** In a monoculture, just one type of plant is planted in a large area. As we discussed earlier, this has proved to be an unwise practice for numerous reasons. **Plantation farming,** which is practiced mainly in tropical developing nations, is a type of industrialized agriculture in which a monoculture cash crop such as bananas, coffee, or vegetables, is grown and then exported to developed nations.

Soil Degradation

Have you ever read *The Grapes of Wrath* by John Steinbeck? Well, the story in this book took place in the 1930s, when droughts in the Great Plains reduced the area to a giant Dust Bowl. Although the drought was the major cause of the Dust Bowl, farming practices used at that time also contributed to the destruction of the land.

In an effort to address the Dust Bowl and other agricultural problems, the United States Soil and Conservation Service (today it's called the National Resources Conservation Service) was established, and it passed the Soil Conservation Act in 1935. Conservation districts were set up by the Service and these franchises provided education to farmers.

Today, farmers can protect soil from degradation in numerous ways. The practice of **contour plowing,** in which rows of crops are plowed across the hillside, prevents the erosion that can occur when rows are cut up and down on a slope. **Terracing** also aids in preventing soil erosion on steep slopes. Terraces are flat platforms that are cut into the hillside to provide a level planting surface; this reduces the soil runoff from the slope. Additionally, **no-till methods** are quite beneficial; in no-till agriculture, farmers plant seeds without using a plow to turn the soil. Soil loses most of it carbon content during plowing, which releases carbon dioxide gas into the atmosphere. (And as you know, increased levels of CO_2 in the atmosphere have been associated with global climate change!)

Finally, **crop rotation** can provide soils with nutrients when legumes are part of the cycle of crops in an area. An alternate to crop rotation is **intercropping** (also called **strip cropping**), which is the practice of planting bands of different crops across a hillside. This type of planting can also prevent some erosion by creating an extensive network of roots. As you might be aware, plant roots hold the soil in place and reduce or prevent soil erosion.

The Livestock Business

Perhaps not surprisingly, the introduction of all these new agricultural techniques has significantly affected the livestock business, and this brings us full circle to the concept of the Tragedy of the Commons. As long as the grazing area is sufficient for the number of animals, livestock grazing is a sustainable practice. If, however, grass is consumed by animals at a faster rate than it can re-grow, land is considered **overgrazed.** Overgrazing is harmful to the soil because it leads to erosion and soil compaction. One solution to the problem of overgrazing is similar to crop rotation—animals can be rotated from site to site and away from their source of water. Another solution involves the overall control of herd numbers.

Various tracts of public lands are available for use as rangeland, and cooperation between government agents, environmentalists, and ranchers can help avoid problems of overgrazing on these lands. The Bureau of Land Management is responsible for managing federal rangelands.

Another problem that arises from the large number of grazing animals worldwide is the large amount of animal waste produced. Manure is not used as fertilizer due to difficulty with transport. It has instead become the most widespread source of water pollution in the United States. Grazing animals also consume 70 percent of the total grain crop consumed in the United States, making them expensive food stuff.

FOREST RESOURCES

Many environmentalists are concerned about the deforestation that is taking place in North America. It is interesting to note that the number of trees growing in North America is approximately the same as 100 years ago, but only 5 percent of the original forests are left. The numbers are approximately the same because of the number of trees growing in national parks and tree plantations. What does this mean? It means that most of the trees in North America are young, and that most forests have been harvested and replanted, and have undergone significant successsion.

Deforestation

Deforestation, or the removal of trees for agricultural purposes or purposes of exportation, is a major issue for conservationists and environmentalists. Worldwide, industrialized countries have a higher demand for wood and less deforestation, while developing countries exhibit a smaller demand for wood, but more deforestation. This can be partly explained by the fact that the deforestation that occurs in developing countries primarily takes place because land is being cleared for pastures and farms. Industrialized countries can also import lumber from developing countries.

Nearly all of the deforestation that takes place in North America is done in order to create space for homes and agricultural plots. In sites where deforestation is occurring, the impact on resident ecosystems is significant. Take, for instance, Canada's Vancouver Island. On this island, whole mountainsides have been stripped bare of the centuries-old forests that once existed. While the lumber industry tries to offset this destruction by planting new trees, the saplings, which won't be harvestable for another 50 years, are no substitute for forests of 300-foot giant redwoods. Remember how we talked about ecological succession in Chapter 5? Where do you think all of the plants and animals that relied upon this forest ecosystem (which was a climax community) went to live?

Despite the moral questionability of this habitat destruction, the lumber industry will not be asked to leave Vancouver Island. This is because it's the island's most important source of income. Fifty cents of every dollar the island earns comes from lumbering—this number easily beats the island's income from tourism, which is the runner-up.

Another environmentally negative by-product of deforestation is seen in countries with tropical forests. In these forests, when trees are removed and farms are placed in the cleared land, the already-poor soil is further degraded and the area can only support crops for a short time. Usually, once the soil will no longer support a crop, the land will be used for grazing, but the soil becomes more and more depleted over time until it has no use for humans.

The negative repercussions of clearing tropical rainforests—the losses in biodiversity, and the erosion and depletion of nutrients in the soil—seem to outweigh the economic gains in many people's opinions. However, for those who would like to take a stand by refusing to purchase wood from tropical rainforests, it is often difficult to determine which wood products come from tropical rainforests and which come from sustainable forests. Various organizations, such as the nonprofit group the Forest Stewardship Council, have developed certifying procedures based on standards that will encourage only the use of the wood from sustainable forests.

How Can We Use Forests Sustainably?

There are three major types of forests, and woods are categorized based on the age and structure of their trees. An **old growth forest** is one that has never been cut; these forests have not been seriously disturbed for several hundred years. Not surprisingly, the controversies that revolve around the issue of deforestation are primarily centered on instances in which deforestation is occurring in old growth forests. As we mentioned in the last section, old growth forests contain incredible biodiversity, with myriad habitats and highly evolved, intricate niches for a multitude of organisms. **Second growth forests** are areas where cutting has occurred and a new, younger forest has arisen naturally. About 95 percent of the world's forests are naturally occurring, and the remaining forests are known as **plantations** or **tree farms.** Plantations are planted and managed tracts of trees of the same age (because they were planted by humans at the same time) that are harvested for commercial use.

It makes sense that those in the forestry business would be concerned about finding a way to promote sustainable forestry, because without forests they have no way of perpetuating their income. From an economic viewpoint, the forest must be managed to continually supply humans' need for wood. The management of forest plantations for the purpose of harvesting timber is called **silviculture.** This relatively modern field has a basic tenet to create a sustainable yield; to do this, humans must harvest only as many trees as they can replace through planting. There are two basic management plans that attempt to uphold this tenet.

- **Clear-cutting** is the removal of all of the trees in an area. This is typically done in areas that support fast growing trees, such as pines. Obviously, this is the most efficient way for humans to harvest the trees, but it has major impacts on the habitat, as in our previous example of Vancouver Island.

- **Selective cutting** is the removal of select trees in an area. This leaves the majority of the habitat in place and has less of an impact on the ecosystem. When selective cutting is used, it's quite difficult to remove these trees from the forest, though. This type of **uneven-aged management** is more common in areas with trees that take longer to grow, or if the forester is only interested in one or more specific types of trees that grow in the area. Another type of uneven-aged management occurs in **shelter-wood cutting**. For **shelter-wood cutting,** mature trees are cut over a period of time (usually 10–20 years); this leaves some mature trees in place to reseed the forest.

In the case of **agroforestry,** trees and crops are planted together. This creates a mutualistic symbiotic relationship between the trees and crops—the trees create habitats for animals that prey upon the pests that harm crops, and their roots also stabilize and enrich the soil.

National Forest Policy

The federal government owns about 35 percent of all land in the United States. The need to preserve some of the land was recognized by President Lincoln when he set aside a park in Yosemite, California as as land grant (the precursor to the National Park System). In 1916 the *National Park System* was created in part to manage and preserve forests and grasslands. Today, in addition to the National Park System, there are several ways the federal government controls forested land: *Wilderness Preservation Areas* are open only for recreational activities with no logging permitted. The *National Forest System*, *Natural Resource Lands*, and *National Wildlife Refuges* are the other groups of federally controlled lands that allow logging with a permit.

You should be aware of two important laws that relate to our federal government's policies on preserving public lands; these are shown below.

Date	Name of Law	What It Does
1964	Wilderness Act	Established a review of road-free areas of 5,000 acres or more and islands within the National Wildlife Refuges or the National Park System for inclusion in the National Preservation System. This act restricted activities in these areas.
1968	Wild and Scenic Rivers Act	Established a National Wild and Scenic Rivers System for the protection of rivers with important scenic, recreational, fish and wildlife, and other values.

One more point about managing treed areas: Recent times have seen an increase in the number of greenbelts, nationally. **Greenbelts** are open or forested areas built at the outer edge of a city. Since no one is permitted to build in them, they can increase the quality of life for people living nearby. They also border cities, putting limits on their growth. Sometimes satellite towns are built outside the greenbelt and interconnected with the city by highways and mass transportation methods; in this way, we can add green spaces in urban areas.

Natural Events (That Create Problems for Humans) in Forests

Certain tree diseases and the existence of pests in trees are two natural problems in forested areas. These can create problems for humans (in addition to the trees) because oftentimes they affect the quality of the food and the number of trees that are available for use. Humans manage these natural events in many different ways: by removing infected trees; by removing select trees or planting them sparsely to provide adequate spacing between them; by using chemical and natural pest controls; by carefully inspecting imported trees and tree products; and by developing pest- and disease-resistant species of trees through genetic engineering.

Forest fires are another natural occurrence. There are three major types of fires that occur in forests, and you should be familiar with them for the test.

- **Surface fires** typically burn only the forests' underbrush and do little damage to mature trees. These fires actually serve to protect the forest from more harmful fires by removing underbrush and dead materials that would burn quickly and at high temperatures, escalating more severe fires.

- **Crown fires** may start on the ground or in the canopies of forests that have not experienced recent surface fires. They spread quickly and are characterized by high temperatures because they consume underbrush and dead material on the forest floor. These fires are a huge threat to wildlife, human life, and property.

- **Ground fires** are smoldering fires that take place in bogs or swamps and can burn underground for days or weeks. Originating from surface fires, ground fires are difficult to detect and extinguish.

One final note about forest fires: Most people believe that forest fires are a bad thing despite the fact that they are part of the natural life of a forest. Some trees and plants even need fire in order for their seeds to germinate. The U.S. Forest Service started an advertising campaign to warn people about the ravages of fires, and soon adopted "Smokey the Bear" to help get the message out. This policy reduced the amount of fires, but it also created conditions for more destructive fires. Under natural conditions, fires burn every few years and consume the fuel (dry leaves, needles, and wood) on the forest floor. However, if there are fewer fires, the amount of fuel can build up to very high levels. When this large amount of fuel ignites, the fires are much hotter and the flames much larger, causing more damage than if the fuel supplies were low. One way to solve the fuel buildup issue is to implement "**controlled burns**." These are small fires started when the conditions are just right and which lower the amounts of fuel. This practice is quite controversial.

Have you got all that information about forests? Let's move on to another vast resource that exists on Earth—our oceans.

OCEAN RESOURCES

The term **fishery** is used in several ways, but its main definition is: The industry or occupation devoted to the catching, processing, or selling of fish, shellfish, or other aquatic animals. In the economic sense, a fishery is the sum of all activities on a given marine resource.

Worldwide, about one billion people depend on fish as their main source of food, and about one million people are currently employed in the fishing industry. Incredibly, about 125 million tons of fish are harvested each year—approximately 75 percent of this total amount is consumed as food by humans, and the other 25 percent is used for other purposes.

For many years, nations were subject to what is known as the 12-mile limit—this limited each nation's territorial waters to just 12 miles from shore. However, in the late 1960s, the depletion of a number of offshore fisheries inspired the United Nations to host a series of international conferences to address the problems of fish scarcity. At this time, the depletion of marine fisheries began to be compared to the Tragedy of the Commons and it was nicknamed the Tragedy of Free Access. The result of this conference was that nations were authorized to extend their limits of jurisdiction to 200 miles from shore.

Today, fishermen must go farther and farther out to sea to catch fish, and need to rely on more sophisticated methods for finding them. Sonar mapping, thermal sensing, and satellite navigation are just a few of the advances that have aided fisherman as fish become scarcer and harder to locate.

By-Catch and Overfishing

Most of the fish that are harvested worldwide come from **capture fisheries;** they are caught in the wild and not raised in captivity for consumption. Some of the techniques that have been developed in order to improve fishing yields are creating problems that relate to overfishing. One of these problems is known as by-catch. **By-catch** refers to any other species of fish, mammals, or birds that are caught that are not the target fish. Some fishing methods that result in by-catch are the use of **driftnets,** which float through the water and indiscriminately catch everything in their path; **long lining,** which is the use of long lines that have baited hooks and will be taken by numerous aquatic organisms; and **bottom trawling,** in which the ocean floor is literally scraped by heavy nets that smash everything in their path, including whole marine mountains known as seamounts. Some advances that have been made in the fishing industry in an attempt to mitigate the problems of by-catch are: restrictions on the use of driftnets, the installation of ribbons on bait hooks that scare away birds and prevent them from being caught, and bans on bottom trawling.

How Many Fish Are Left?

It has recently been reported that about 47–50 percent of the major fish stocks of the world are fully exploited. Close to another 20 percent of the stocks are nearly overexploited, and about 10 percent are depleted; and this is mostly due to overfishing.

One partial solution to the problem of overfishing is **aquaculture,** which is the raising of fish and other aquatic species in captivity for harvest. In general, the fish that are raised in captivity are those with the highest economic value—for example, salmon and shrimp. Various different methods are used in aquaculture—some fish are raised totally in captivity and then harvested, while others (like salmon) are initially hatched in captivity, but then released into the wild and captured later. Some saltwater aquaculture is performed in shallow coastal areas; though this is generally for the raising of seaplants and mollusks.

While aquaculture, also known as **fish farming**, does help to meet worldwide demands for fish, it is not a panacea for all of our fishery problems. One concern about aquaculture is the possibility of the accidental release of farmed fish into the wild, which has the potential to introduce new diseases to ocean fish and contaminate the native gene pool. Carp release from fish farming is currently causing issues in the Mississippi River and the Great Lakes. Another problem lies in the fact that many fish that are raised in captivity are carnivorous and are fed captured wild fish, which defeats the purpose of the attempt to kill fewer wild fish!

Most of the public outcry about the endangered animals of the sea has centered on two groups: the dolphins and the whales. Dolphins are a high-profile by-catch, and as you may have noticed, many cans of tuna now advertise as having been caught using "dolphin safe" nets. The slogan "Save the Dolphins" has been frequently employed by international marine conservation groups. However, that slogan is impossible to obey unless humans work to save the natural *habitat* of these creatures, first.

The International Whaling Commission (1974) regulates whaling. Recent policies implemented by the IWC allow the capture of a certain number of whales annually—by Norway for human consumption and by Japan for scientific use. It has recently come to light, however, that Japan has been eating the whales it catches, and they have stated that their rationale is that whales eat too many fish that could instead be caught by humans. Another industry that has recently been criticized for damaging whales' ecosystems is the tourism industry—whale watching tours are said to disrupt whale migration patterns and cause the whales undue stress.

Two Endangered Aquatic Ecosystems

As we've reviewed in earlier chapters, coral reefs are structures found in warm, shallow tropical waters that represent diverse and ecologically crucial ecosystems. Coral reefs are created by small marine animals (called **cnidarians**), which are involved in mutualistic relationships with photosynthetic algae called **zooxanthellae.** Reefs provide local populations with a great variety of seafood, and are also important recreational areas for humans.

In many areas of the world, exploitation has led to severe and irreversible damages to these reefs. One example of such irreversible coral damage is **coral bleaching**. In coral bleaching, higher-than-usual water temperatures cause the death of the zooxanthellae, and this in turn causes the death of the coral reef. While some bleaching is normal, high water temperatures can be caused by weather fluctuations such as El Niño, and since this is a periodic event, coral bleaching is an ongoing concern.

Another threatened aquatic ecosystem is the mangrove swamp. Mangrove swamps are coastal wetlands found in tropical and subtropical regions, and they are threatened by activities such as shrimp aquaculture and the degradation of the Western coastlines. Mangroves are characterized by trees, shrubs, and other plants that can grow in brackish tidal waters, and are often located in estuaries, which as you learned earlier, are areas where freshwater meets salt water. In North America, mangrove swamps are found from the southern tip of Florida along the entire Gulf Coast to Texas; Florida's southwest coast supports one of the largest mangrove swamps in the world.

A huge diversity of animals is found in mangrove swamps. Because these estuarine swamps are constantly replenished with nutrients transported by freshwater runoff from the land, they support a bursting population of bacteria, other decomposers, and filter feeders. These ecosystems also sustain billions of worms, protozoa, barnacles, oysters, and other invertebrates, which in turn feed fish and shrimp, which support wading birds, pelicans, and in the United States, the endangered crocodile.

The importance of mangrove swamps has been well established. They function as nurseries for shrimp and recreational fisheries, exporters of organic matter to adjacent coastal food chains, and enormous sources of valuable nutrients. Their physical stability also helps to prevent shoreline erosion, shielding inland areas from severe damage during hurricanes and tidal waves.

Along with the Whaling Commission, there are many laws and regulations pertaining to preserving ocean resources. A few that the College Board is likely to ask about are the following:

Date	Name of Legislation	What It Did
1965	Anadromous Fish Conservation Act	Protected fish that live in the sea but grow up and breed in fresh water
1976	Magnuson Fishery Conservation and Management Act	Governed the conservation and management of ocean fishing
1972	Marine Mammal Protection Act	Established a federal responsibility to conserve marine mammals
1973	Endangered Species Act	Provided broad protection for species of fish, wildlife, and plants that are listed as threatened or endangered in the U.S. or elsewhere

Finally, there are some international agreements.

Date	Name of Legislation	What It Did
1982	The United Nations Agreement for the Implementation of the Provisions of the United Nations Convention on the Law of the Sea	Set out the principles for the conservation and management of certain types of fish
1975	CITES (the Convention on International Trade in Endangered Species of Wild Fauna and Flora)	An international agreement between governments that ensured that international trade in specimens of wild animals and plants do not threaten their survival

We're done with our discussion of resources from the sea. Let's move on to talking about resources that come from underground.

MINING

Mining is the excavation of earth for the purpose of extracting ore or minerals. We can divide mineral resources into two main groups by how they're used. **Metallic minerals** are mined for their metals (for example, zinc), which can be extracted through smelting, and used for various purposes. **Nonmetallic minerals** are mined to be used in their natural state—nothing is extracted from them. Examples of nonmetallic minerals are salt and precious gems. Here's one more term you should know for the exam, if you don't already: A **mineral deposit** is an area in which a particular mineral is concentrated. An **ore** is a rock or mineral from which a valuable substance can be extracted at a profit.

The cost of extracting minerals depends on numerous factors, including the location and size of the mineral deposit. Additionally, the impetus for mining certain deposits more than others is often purely based on the value of the mineral resource. Understandably, the higher the value of the resource, the more money and effort will be put into mining it.

Environmental concerns about mining do not center on the depletion of mineral resources from the earth's surface. Instead, they revolve around the damage that is done during the extraction process. The extraction of a mineral from Earth generally disrupts the ecosystem and scars the land. Sometimes the extraction leaves pollutants that were associated with the mineral ore underground or the machinery used to extract it. One example of this is the deposition of iron pyrite and sulfur in the mining of coal. The acid forms as water seeps through mines and carries off sulfur-containing compounds. The chemical conversion of sulfur-bearing minerals occurs through a combination of biological (bacterial) and inorganic chemical reactions; and the result is the buildup of extremely acidic compounds in the soil surrounding the deposit. These compounds create acid mine drainage that can severely harm local stream ecosystems. In mining processes, waste material is called **gangue,** and piles of gangues are called **tailings.**

One of the most controversial types of mining is **strip mining**, which involves stripping the surface layer of soil and rock (this layer is called **overburden**, and it is whatever material lies above an area of scientific interest) in order to expose a seam of mineral ore. This type of mining is only practical when the ore is relatively close to the surface, which is why it's used mainly for coal mining. This is the least expensive—and least dangerous—method of mining for coal. However, because strip mining requires removing massive amounts of top soil, it has a much greater impact on the surrounding environment than underground mining. The most extreme form of strip mining, mountaintop removal, transforms the summits of mountains and destroys ecosystems. This method is mostly associated with coal mining in the Appalachian Mountains. With **shaft mining**, vertical tunnels are built to access and then excavate minerals that are underground and otherwise unreachable.

Another environmental drawback to mining is that the refinement of these minerals often requires extensive energy input. For example, it takes approximately 15.7 kW of electricity to produce one kilogram of pure aluminum from its ore. On the other hand, recycling aluminum requires only 5 percent of the energy that's required to smelt it, and generates only 5 percent of the greenhouse gases. Recycle those soda cans!

After minerals have been extracted from their ore, they may be used in their rough form or further processed. Aluminum, for example, must be further refined after it is mined. Coal is an exception. After mining, it is transported to a power plant and burned in its original state. Sometimes two metals are combined to form a product; this is the case with stainless steel, which is a combination of iron and either nickel or chromium, and regular steel, which is 95.5 percent iron and 0.5 percent carbon. Because of the energy expended in mining and extraction, the steel industry is responsible for much of the air pollution that exists today!

Fortunately, air, land, and water harmed by mining can be reclaimed through **mine restoration** projects. In 1977, Congress passed the Surface Mining Control and Reclamation Act (SMCRA), which created one program to help coal mines manage pollutants and another to guide the reclamation of abandoned mines.

Mineral Production

The following table shows you the production (in thousands of metric tons) of some non-fuel mineral resources. While you will not have to memorize the amounts for the exam, you can appreciate how much is produced. Remember that these high production rates lead to the eventual depletion of these resources. Also, be ready to describe the impact of mining and mineral production on the environment.

Mineral	2004 Production (Thousands of metric tons)
Copper	14,600
Phosphate rock (for fertilizer)	141,000
Bauxite (aluminum ore)	159,000
Iron ore	1,340,000
Zinc	9,600,000

As you can see from the chart, demand for mineral resources is very high. The increased demand for manufactured goods means that we need to extract more and more raw materials from the earth. The chart below, from the U.S. Geological Service, shows the number of years of supply for five selected mineral ores. For the exam, you should be aware of the need to use mineral resources in a sustainable manner.

Mineral	Years of Supply
Bauxite (aluminum ore)	152
Copper	32
Iron Ore	105
Nickel	41
Zinc	22

Finally, there are several laws that govern mining in the United States. These laws deal with the exploration and mining of minerals and the negative impacts of waste and pollution that result from mining.

Date	Name of Legislation	What It Did
1872	Mining Act	Governed prospecting and mining of minerals on publicly owned land
1920	Mineral Leasing Act	Permitted the Bureau of Land Management to grant leases for development of deposits of coal, phosphate, potash, sodium, sulphur, and other leasable minerals on public domain lands
1980	Comprehensive Environmental Response, Compensation, and Liability Act (Superfund)	Regulated damage done by mining
1976	Resource Conservation and Recovery Acts (RCRA)	Regulated some mineral processing wastes
1977	Surface Mining Control And Reclamation Act	Established a program for regulating surface coal mining and reclamation activities. It established mandatory standards for these activities on state and federal lands, including a requirement that adverse impacts on fish, wildlife, and related environmental values be minimized.

ECONOMICS AND RESOURCE UTILIZATION

The study of how people use limited resources to satisfy their wants and needs is called economics. As you can imagine, some of those needs are tangible (food and shelter are two examples), while others are intangible (the beauty of a forest and clean air are two examples). A resource can have both tangible and intangible properties.

A forest has value for supplying jobs and wood (tangible), as well as for its beauty and ability to remove CO_2 from the air (intangible). When private citizens, governments, and corporations make a decision on how to use the forest, they must weigh the benefits (more jobs or lumber) against the cost of cutting down the trees (less recreation space, the loss of biodiversity, or decrease in CO_2 removal). This process is called **cost-benefit analysis**. It may be easy to assign a monetary value to the tangible properties (like the amount of lumber in the forest), but how do you assign a monetary value to the intangible properties (like the beauty of a forest)? While cost-benefit analysis helps make decisions on how to use resources, you can see that the process is very difficult, and it can lead to different estimates by different groups.

Economists also want to figure out the cost of each step in a process. From our forest example, what is the cost to the economy of removing one more acre to the forest; or what is the benefit to us if we add one more acre to the forest? The additional costs are termed **marginal costs**; the added benefits are called **marginal benefits**. It is important to remember that resources are not free and unlimited. Some resources must be expended in order for us to use them. While we may benefit from more acres to hike in, the lumber company will suffer from not having as many trees to cut. In other words, marginal benefits and costs help us understand tradeoffs. By preserving a forest, we trade more hiking space with less profit for local economies.

As we use resources, there are often unwanted or unanticipated consequences of our using those resources, or **externalities**. These can be positive, when the result is good, and negative, when the result is bad for the environment. Consider buying a television, for example. When you buy the television there are costs—you pay for the labor, raw materials and electricity to run it. After you buy the television, the dealer uses some of that money to pay employees to clean up litter on a highway. That cleanup benefits everyone (positive externalities), even those who did not buy the television. On the other hand, there are also negative externalities. If you watch a lot of television, you use a lot of electricity. That electricity is generated by burning coal, and that causes acid rain. The damage done by the acid rain harms everyone—a negative externality.

One final thing to remember: the use of economics (cost-benefit analysis, marginal costs and benefits, and externalities) to make choices dealing with environmental issues is morally neutral. These economic factors do not say anything about the ethics or fairness of those choices. There are situations in which we make decisions not based on the best balance between marginal costs and benefits, but on what is best for everyone. Take water pollution, for example. If a toxic chemical "X" is in a stream, there is a cost to clean it up. We make the decision to clean up most of chemical "X," even if the marginal costs exceed the marginal benefits, because removing chemical "X" will keep all of the people healthy.

We're done discussing resources. As we've alluded to many times, as populations increase, more pressure is placed on Earth's natural resources, and along with this comes the need for humans to find ways to develop those natural resources for direct human use. With this in mind, let's move on to the next chapter and review energy resources and consumption.

KEY TERMS

Here are your key terms for Chapter 7. Know what you should do with them? No, don't skip them! Learn them!

Resources
resource
conservation
preservation
natural resources
ecosystem capital
renewable resources
nonrenewable resources
consumption
production

Agriculture
subsistence agriculture
slash and burn agriculture
Green Revolution
FIFRA
salinization
integrated pest management
genetically modified organism
monoculture
photosynthate
plantation farming
contour plowing
terracing
no-till methods
crop rotation
inter cropping
strip cropping
overgrazing

Forestry
deforestation
old growth forest
second growth forest
plantations
silviculture
clear-cutting
selective cutting
uneven-aged management
shelter-wood cutting
greenbelts
agroforestry
surface, crown, ground fires

Oceans
fishery
capture fisheries
by-catch
driftnets
long lining
bottom trawling
aquaculture
fish farming
cnidarians
zooxanthellae
coral bleaching

Mining
ore
metallic and nonmetallic minerals
mineral deposit
gangue
tailings
strip mining
overburden
mountaintop removal
shaft mining
restoration

Economics
cost-benefit analysis
marginal costs and benefits
externalities

Chapter 7 Drill

Directions: Each of the questions or incomplete statements below is followed by five suggested answers or completions. Select the one that is best in each case. For answers and explanations, see Chapter 13.

1. Which of the following correctly describes the process of smelting?

    (A) Separating the desired metal from other elements in the ore
    (B) Cleaning up drainage from mines
    (C) Detoxifying harmful chemicals
    (D) Removing ore from underground mines
    (E) Making gasoline

2. In a very polluted river it costs $3 per kilogram to remove the first 80% of the pollution. It costs $25 per kilogram to remove the last 20% of the pollutant. This phenomenon is correctly referred to as:

    (A) cost-benefit analysis
    (B) external costs
    (C) marginal costs
    (D) marginal benefit
    (E) externalities

3. Which of the following correctly describes the process of clear-cutting?

    (A) Some mature trees are left to provide shade for younger trees.
    (B) Only trees with commercial value are cut down.
    (C) A few mature trees are left to reseed the land after cutting.
    (D) All the commercially usable trees in an area are cut down.
    (E) Trees are planted between rows of other crops.

4. Moderate irrigation with groundwater over a long period of time can cause:

    (A) salinization
    (B) waterlogging
    (C) desertification
    (D) succession
    (E) leeching of minerals from the soil

5. All of the following are problems created by the deforestation of rainforests EXCEPT

    (A) increased erosion
    (B) loss of biodiversity in the area
    (C) changes in local rainfall levels
    (D) an increase in the availability of grazing land
    (E) loss of soil fertility

6. Greenbelts are useful to

    (A) slow the process of urban growth
    (B) get more crops out of farmland
    (C) maintain borders around a person's home property
    (D) prevent erosion
    (E) hide unwanted objects from people's view

7. Which of the following government agencies is responsible for the management of federal rangeland?

    (A) The U.S. Park Service
    (B) The U.S. Bureau of Mines
    (C) The Bureau of Land Management
    (D) The Environmental Protection Agency
    (E) The U.S. Commerce Department

8. Which of the following is NOT a renewable resource?

    (A) Air
    (B) Soil
    (C) Copper ore
    (D) Water
    (E) Biodiversity

9. Nations have overfished international waters and have depleted many commercially important fish species. This is a good example of which of the following?

    (A) International agreements
    (B) The Tragedy of the Commons
    (C) The Rule of 70
    (D) Trade barriers
    (E) Sustainability

10. Which of the following best describes industrialized agriculture?

    (A) Consumes large amounts of fossil fuels, pesticides, and water
    (B) Uses human labor and draft animals to grow crops
    (C) Rows of crop plants are interspersed with rows of trees
    (D) Uses little water or fossil fuels; relies on human labor
    (E) Crops are grown on small plots of land

11. The international trade in endangered species is regulated by which of the following?

(A) The Endangered Species Act
(B) Marine Mammal Protection Act
(C) The National Environmental Policy Act
(D) RCRA
(E) CITES

12. Which of the following are problems that have emerged with the overuse of pesticides?

I. Better crop yield
II. Pesticide-resistant pests
III. Improved human health

(A) I only
(B) II only
(C) III only
(D) I and III only
(E) I, II, and III

13. Which of the following is true concerning the use of National Parks?

(A) They can be used for cutting timber as well as recreation.
(B) They can be used for mining as well as recreation.
(C) They can be used only for camping, fishing, and boating.
(D) They can be used for conservation of natural habitat
    as well as livestock grazing.
(E) They can be used for military activities and the development of natural gas reserves.

14. The acid most commonly found in mine drainage is

(A) carbonic acid
(B) sulfuric acid
(C) hydrochloric acid
(D) acetic acid
(E) citric acid

15. The World Trade Organization strives to

(A) protect endangered species on land
(B) regulate the global fishing industry
(C) move toward the globalization of all
    the nations
(D) establish rules for the free flow of economic
    goods and services between countries
(E) decrease competition for goods among nations

Directions: Each set of lettered choices below refers to the numbered questions or statements immediately following it. Select the one lettered choice that best answers each question or best fits each statement. A choice may be used once, more than once, or not at all in each set.

Questions 16-19 refer to the various methods used to catch fish.

    (A) drift net
    (B) long-line fishing
    (C) aquaculture
    (D) bottom trawling

16. A weighted net is dragged across the sea floor

17. Marine organisms are raised in a bay or confined area

18. This floats through the water indiscriminately catches everything in its path

19. Baited hooks attached to lines are dropped off the side and then reeled back onboard

Free-Response Question

1. The irrigation of farmland is vital to the production of the world's food supply. In China, 87 percent of the water withdrawn is used for irrigation. In the United States, this figure approaches 41 percent. Most of the water is applied to the land in a process called gravity irrigation, in which the water is simply allowed to flow, via the force of gravity, into the fields.

   (a) Describe one positive and one negative impact of gravity irrigation.
   (b) Describe one alternative to gravity irrigation. Give one positive and one negative effect of that practice.
   (c) Massive irrigation programs can also impact underground water supplies. Describe one negative impact that irrigation might have on those supplies.
   (d) Dams are often used to create irrigation water reservoirs. Describe two positive and two negative impacts that a large dam would have on the immediate area around it.

REFLECT ACTIVITY

Respond to the following questions:

- For which content topics discussed in this chapter do you feel you have achieved sufficient mastery to answer multiple-choice questions correctly?

- For which content topics discussed in this chapter do you feel you have achieved sufficient mastery to identify hot-button topics and successfully articulate them in free-response questions?

- For which content topics discussed in this chapter do you feel you need more work before you can answer multiple-choice questions correctly?

- For which content topics discussed in this chapter do you feel you need more work before you can successfully identify and articulate them in free-response questions?

- What parts of this chapter are you going to re-review?

- Will you seek further help, outside of this book (such as a teacher, tutor, or AP Central), on any of the content in this chapter—and, if so, on what content?

Chapter 8
Energy

For the AP Environmental Science Exam, you'll be expected to know the basics of the concept of energy; in fact, according to the AP outline, 10–15 percent of the questions you'll see on the exam relate to energy.

Unlike the essential elements we discussed in earlier chapters, energy flows on a one-way path through the atmosphere, hydrosphere, and biosphere, and is essential for living organisms in many of its forms. At the most fundamental level, **energy** is defined as the capacity to do work. There are three types of energy: **potential energy** is energy at rest—it's stored energy—while **kinetic energy** is energy in motion. You might recall from your physics class that potential energy can be converted to kinetic energy. The third type of energy is in the form of **radiant energy,** or sunlight. Two terms that describe the movement of energy around Earth are **convection,** which is the transfer of heat by the movement of the heated matter, and **conduction,** which is the transfer of energy through matter from particle to particle. Keep in mind that different energy sources are capable of storing types of energy that differ in quality. For example, both wood and coal will burn to produce heat, but coal produces more heat because it contains higher **energy quality**. **Net energy yield** refers to the comparison between the cost of extraction, processing, and transportation, and the amount of useful energy derived from the fuel.

As you saw in our chapter about weather, convection and conduction are very important processes that drive the movement of water in the hydrosphere of Earth. This is just one way in which the flow of energy around the earth affects every process—geological or biological—that takes place.

In this chapter, we'll begin with a discussion of the basic units of energy, and go through the two laws of thermodynamics that you'll need to know for the test. After that, we'll begin our discussion of Earth's energy resources—the earth provides humans with resources of energy, just as it does the physical, natural resources that we learned about in the last chapter. Let's begin!

UNITS OF ENERGY

For this exam, you'll be expected to recognize the following units of energy and power:

- **Energy Units:** Joule (J), Calorie (cal), British thermal unit (Btu), and kilowatt hour (kWh), which is a measure of watt/time

- **Power Units:** Watt (W) and Horsepower (hp)

Remember, a watt is equal to volts × amperage. You should also be intimately familiar with the First and Second Laws of Thermodynamics, so let's review those before we move on—and make sure you memorize these before test day!

Laws of Thermodynamics

1. The **First Law of Thermodynamics** says that energy can neither be created nor destroyed; it can only be transferred and transformed. One example of such a transformation occurs in photosynthesis. In photosynthesis, radiant energy from the sun is converted to chemical energy in the form of the bonds that hold together atoms in carbohydrates.

2. The **Second Law of Thermodynamics** says that the entropy (disorder) of the universe is increasing. One corollary of this Second Law of Thermodynamics is the concept that, in most energy transformations, a significant fraction of energy is lost to the Universe as heat; for example, as we reviewed in Chapter 5, in food chains, only about 10 percent of the energy from one trophic level is available for the next energy upon consumption.

Okay, those are the basics about energy that you'll need to know for the test. Now let's begin our review of the energy resources that exist on Earth.

NONRENEWABLE ENERGY

Perhaps surprisingly, one of our biggest uses of energy is in the production of electricity. In other words, we use tons of energy each year to produce electricity—another form of energy!

In general, electricity is produced in the following way: An energy source provides the power that heats up water, transforming it into steam, which then turns a turbine. Hence, the turbine converts kinetic energy (from the steam) into mechanical energy (the spinning of the turbine). Now here's where the generator comes in. The generator consists of copper wire coils and magnets, one of which is stationary (stator) and the other of which rotates (rotor). As the turbine spins, it causes the magnets in the generator to pass over the wire coils (or vice versa), generating a flow of electrons through the copper wire and thus producing an alternating current that passes into electrical transmission lines. In lieu of steam, flowing water or wind can also provide the power needed to turn the turbine and produce electricity.

So, where do we get the energy that we use to heat up that water in the first step of the creation of electricity? Well, the three main sources for the global production of electricity are

- fossil fuels (provide 64 percent of the world's electricity)

- nuclear energy (provides 17 percent of the world's electricity)

- renewable energy sources (provide 19 percent of the world's electricity)

Let's go through each of the types of energy above and see where they came from, what effect their use has on the earth, and how sustainable they are.

Fossil Fuels

During the Industrial Revolution (in the early eighteenth century), steam was produced almost exclusively through the burning of firewood and coal—and this, in turn, provided the energy for most mechanical processes. Today, oil is our primary power source. About 35 percent of total global energy production comes from oil products; the runner-up to oil is coal, and the runner-up to coal is natural gas. Together these three fossil fuels provide 80 percent of the world's energy.

Fossil fuels, as the name indicates, are formed from the fossilized remains of once-living organisms. Over vast tracts of time, this organic matter was exposed to intense heat and pressure. Eventually, theses factors broke down the organic molecules into oil, coal, and natural gas.

Oil, or petroleum, is made of long chains of hydrocarbons; and coal contains a mixture of carbon, hydrogen, oxygen, and other atoms. Natural gas is made mostly of methane gas (CH_4) with a mixture of other gases.

Generally, oil and natural gas are formed in the same areas. These materials are found deep in the earth under both land and ocean floor, where they are stored in the pores (spaces) between rocks. Coal is found in long continuous deposits, called **seams,** at various depths underground. The seams represent areas where large amounts of plant remains were buried and eventually transformed into coal. We will cover the process of coal mining on the next page.

Certain types of geologists locate fossil fuel reserves. They plan and supervise the extraction of these fuels from the earth. Using knowledge of the geology and rock formations, these scientists make predictions about which sites are most likely to have fossil fuel deposits. They use **exploratory wells** to drill and sample a particular area. If an exploratory well hits a fossil fuel reserve, it can provide an estimate of the amount of fuel that can be obtained from that area; this is called the **proven reserve**. It is important to know that although exploratory wells can provide a fairly precise estimate of the size of a reserve, these numbers are just educated guesses (not so proven after all!). The amount of a resource that can be extracted from a reserve is dependent on the technologies available and the cost of extraction. If extraction costs are too high, it is not economically feasible to extract the resource. If a coal seam is buried very deeply, for instance, it may cost more money and fuel to extract it than the value of the seam.

What About Oil?

When oil is pumped up fresh from a reserve, it is called **crude oil**. Crude oil varies greatly from reserve to reserve. It can range from thin to viscous (thick); from high sulfur to low sulfur; it can even vary in color and odor.

There are three different methods of extracting oil. In primary extraction, the oil can be easily pumped to the surface. When some oil wells are tapped for the first time, there is a large release of oil and gas, a *gusher,* due to the pressure in the reserve. When the oil is harder to extract, people rely on pressure extraction, which uses mud, saltwater, and even CO_2 to push out the oil from the reserve. The final method utilizes steam, hot water, or hot gases to partially melt very thick crude oil and make it easier to extract. Oil reserves can also be found in rock (shale oil) and surface sands (tar sands).

Drilling for oil is only moderately damaging to the environment because little land is needed to drill. However, since oil is transported thousands of miles by tankers, pipelines, and trucks, a lot of environmental damage can occur during transportation.

In 2010, the United States suffered one of its worst environmental disasters when an explosion on the Deep Water Horizon drilling rig caused oil to spill from the well into the Gulf of Mexico for three months. This was the largest marine oil spill in the history of the oil industry. Eleven men were killed during the explosion, and the spill caused a great deal of damage to both marine and wildlife habitats and local economies along the coast.

What About Coal?

Let's spend some time reviewing coal. The qualities of different types of coal are ranked by the number of BTUs that they produce upon burning. The purest coal is called **anthracite**; this coal is almost pure carbon. The second-purest coal is **bituminous,** followed by **subbitumimous,** and finally **lignite**—the least pure coal. Coal mining occurs through one of two processes—strip mining or underground mining—both of which can be hazardous and have serious environmental impacts. **Underground mining** involves the sinking of shafts to reach underground deposits. In this type of mining, networks of tunnels are dug or blasted and humans enter these tunnels to manually retrieve the coal. After production stops at these mines, cave-ins can occur, causing massive slumping or **subsidence. Strip mining** involves the removal of the earth's surface, all the way down to the level of the coal seam. The coal is then removed and then the **overburden** (the earth that was removed) is replaced, topped with soil, and the area is contoured and re-vegetated. Most states require strip mine owners and operators to completely reclaim areas that are mined by taking all of the steps outlined above. However, the process of mining and removing the coal from the earth leaves hazardous slag heaps containing sulfur that can be leached out and enter the water table.

At this time, coal is the most abundant fossil fuel, and it is used to generate electricity in over 50 percent of the power plants in the United States.

The use of coal to produce electricity has several disadvantages; for one, when it is burned in the production of electricity, carbon dioxide, nitrogen oxides, mercury, and sulfur dioxide—all of which contribute to air pollution—are released as byproducts. However, some of these by-products can be removed through the actions of **scrubbers,** which contain alkaline substances that precipitate out much of the sulfur dioxide. The neutral compound formed in the scrubber (calcium sulfate) is eliminated in waste sludge. Two other waste products produced by the burning of coal are **fly ash** and **boiler residue**—you should be familiar with both of these terms for the exam.

Another problem with coal is that it often contains a significant amount of the element sulfur, both in the form of iron sulfide (pyrite) and as organic sulfur. Sulfur is another contributor to air pollution. While iron sulfide can be removed by grinding the coal into small lumps and washing it, organic sulfur is only released during the combustion (burning) of coal. However, scrubbers can remove organic sulfur from the flue gases after the coal is burned. Another solution to this problem is to burn the coal with limestone—the liberated sulfur combines with the calcium in limestone to form calcium sulfate; and this prevents it from being released through the flue. There are different types of smoke-stack scrubbing: wet scrubbing, and baghouse or cyclo scrubbing. Wet scrubbing uses a fine mist of water to transform sulfur oxides (SO_x) from an air pollution issue to either a water pollution issue or to a commercial product, sulfuric acid. This is similar to how a rain storm will cleanse the air of pollen and dust that causes allergies in some people. Baghouse, or cyclo, scrubbers are very similar to large vacuum cleaners that either filter (baghouse) or spin (cyclo) particulates out of the effluent gases. Electrostatic filters use electric charge to attract dust particulates to metal surfaces where they can be gathered and disposed of as solid waste. This is similar to your television screen being the dustiest surface in your living room due to electric properties of the device.

You've probably seen lots of news reports lately about the harmful effects—particularly to babies and very young children—of ingesting seafood that's contaminated with mercury. An EPA study found that one in six women of childbearing age in the U.S. may have blood mercury levels that could be harmful to a developing fetus. Coal-fired power plants are the major source of mercury pollution in the environment. Airborne mercury pollution can deposit in the ground as a result of rainfall, and into lakes, streams, or other bodies of water both directly from rainfall, and as a result of runoff—and the mercury can travel hundreds of miles from its source through the air before being deposited. Mercury in water can accumulate in fish, which are then eaten by people. (Biomagnification is what we call this increase in the concentration of a substance.) Abandoned metal and coal mines frequently produce **acid mine drainage**, highly acidic water which flows to surrounding areas.

What About Natural Gas?

The third fossil fuel you need to know about is natural gas. Natural gas is made mostly of methane (CH_4) as well as pentane, butane, and several other gases in small quantities. As you learned earlier, natural gas is produced by the actions of heat and pressure over long periods of time. In today's environment, it is also produced by living organisms (mostly by anaerobic bacteria). Methane-producing bacteria can be found in landfills, swamps, and the intestines of various animals. Here's an interesting fact: While the largest source of methane is wetlands, the second largest source is our flatulent livestock.

Currently, natural gas is used for home-heating and cooking. It can also be burned to generate electricity. Some power plants are designed to switch between oil and natural gas fuels depending on the cost. The engines of cars and trucks can be modified to burn natural gas instead of gasoline. There is a landfill operator in the state of New Jersey who tested a process of trapping methane from a landfill, liquefying it and then using the liquid methane to power the trucks that bring garbage to the landfill.

Because of its simple molecular structure, natural gas produces only carbon dioxide and water when it burns. It does not produce the oxides of nitrogen and sulfur associated with burning coal or oil. Before you get really excited about natural gas, you should be aware of its dangers. In an uncontrolled release (like a leak), it can cause violent explosions. It is also more difficult to transport than coal or oil. Because a tank can hold a small amount of gas, producers liquefy it by putting the gas under high pressure (**L**iquefied **N**atural **G**as). This process requires energy. Natural gas can also be transported by pipes; pipes carry the risk of leaks and explosions, and some habitats are damaged during the building of the pipe system.

How Much Fossil Fuel Is Left?

In order to understand how long our accessible fossil fuel supplies will last, you should know how quickly we are using up those fuels. Let's take oil, the most widely used fuel, as an example. The table below represents the amount of petroleum (thousands of barrels) that selected countries use each day. These data are from 2012, the last year for which we have this information.

Country	2012 Petroleum Consumption (Millions of barrels per day)
United States	18.5
China	10.3
Japan	4.7
India	3.6
United Kingdom	1.5

As you can see, the United States is by far the largest consumer of petroleum. A quick bit of addition shows that these five countries alone consume almost 38.6 million barrels of oil each day! As you can imagine, that leads some scientists to ask questions about how long our supplies of oil (and the other fossil fuels) will last. One well-known authority on the future of oil production, the late M. King Hubbert, stated that the end of oil as a cheap and easily available form of energy is in the near future, and that we must begin to develop alternative fuel sources. This theory is commonly known as Hubbert Peak.

Nuclear Energy

Nuclear energy is the world's primary non-fossil fuel, nonrenewable energy source. In the United States, 20 percent of electrical energy is provided by nuclear power plants. Worldwide, more than 400 nuclear power plants produce approximately 13 percent of the world's electrical energy. China and India lead the world in the creation of new nuclear facilities.

Let's talk a little bit about what exactly these nuclear power plants do. The current nuclear plant technology involves the use of uranium-238, which is enriched with 3 percent uranium 235. The uranium 235 isotope is split in a process called **fission**; this is the key reaction in the production of nuclear energy. **Breeder reactors** generate new fissionable material faster than they consume such material.

Furthermore, they can use a more abundant form of uranium, uranium-238, or an alternative called thorium.

However, the future of nuclear power will probably involve **nuclear fusion,** which is the process of fusing two nuclei (most likely two isotopes of hydrogen—tritium-2 neutrons and deuterium-1 neutron). One other important thing you should know for the exam is that radioactive materials all have **half-lives,** which is the time it takes for half of the radioactive sample to degrade.

In the Unites States, there are two types of nuclear reactors. They are known as boiling water reactors and pressurized water reactors.

- **Boiling Water Reactors**—These reactors use the heat of the reactor core to boil water into steam. This steam is piped directly to the turbines. The steam spins the turbines that generate the electricity. The water is cooled back to a liquid (by a heat exchanger), then pumped back to the core to be turned into steam again. This reactor uses two water circulations systems; one system makes steam and carries it to the turbine and the other cools the water from the core so it can be turned back into steam.

Boiling Water Reactor

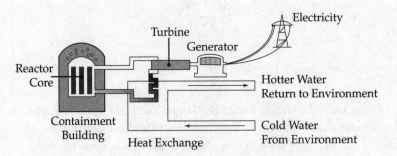

- **Pressurized Water Reactors**—These reactors use the heat from the core to heat a second water supply via a heat exchanger. This second water system provides the steam to spin the turbines. A third water circulation system cools the steam from the turbines (by using a second heat exchanger) so it can be used to make steam again. These reactors use three water circulation systems; the first cools the core, the second makes the steam, and the third cools the steam back into water so it can be made into steam again.

Pressurized Water Reactor

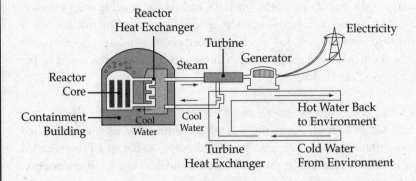

Some arguments in favor of the use of nuclear power include the fact that the production of nuclear energy produces no sulfur dioxide or nitrogen oxide and less carbon dioxide than does the production and combustion of fossil fuels.

Since the first testing of atomic weapons, people have been concerned about safety issues surrounding nuclear energy. The nuclear reactor accident that occurred at the Three Mile Island facility in Pennsylvania in 1979, the devastating explosion that occurred at the Chernobyl facility in the Ukraine in 1986, and the catastrophic failure at the Fukushima nuclear power plant in Japan in 2011 brought some major safety concerns to the public's attention.

The chart below shows a number of these issues.

Safety Issue	Description
Meltdown	Reactor loses coolant water and thus the very hot core melts through the containment building. The radioactive materials could then get into the groundwater.
Explosion	Gases generated by a uncontrolled core burst the containment vessel and spread radioactive materials in the environment.
Nuclear weapons	Some of the by-products of the fission reaction can be remade into fission bombs, or "dirty bombs," that spread damaging radioactive isotopes.
Highly radioactive waste	No longer usable cores, piping, and spent **fuel rods** need to be stored for many centuries. The "spent" fuel can contain radioactive elements like plutonium-239 that has a half-life of 2.13×106 years.
Thermal pollution	The water used to cool turbines is returned to local bodies of water at a much higher temperature than when it was removed unless first cooled.
Radioactive elements	Gamma rays produced by radioactive decay can damage cells and DNA, which can cause breast, thyroid, stomach, and leukemia cancer. Damage to the immune system can also result.
Concern for one's safety	People suffer from mental stress, anxiety, and depression caused by concerns for their safety, resulting in **Not In My Backyard Syndrome (NIMBY)**.

At this point, the cost for building a new nuclear power plant in the United States is prohibitive due to changing regulations; and worldwide, more than 100 nuclear power plants have been decommissioned. The main reason for this is the problem of the nuclear waste that's created. The United States' spent fuel is currently stored on site at nuclear power plants, but the plants are running out of space. The proposed final-destination storage site is Yucca Mountain (in Nevada), within the boundaries of a former nuclear weapons testing site. This storage site is located in a remote desert on federally protected land, but the storage facility will not be in operation until 2017 at the earliest. Also, no insurance company will insure nuclear power plants, which leaves this important issue to the federal government.

RENEWABLE ENERGY

Obviously, the advantage to discovering or developing renewable energy sources lies in the fact that these types of energy sources are bottomless. However, globally only about 14 percent of our energy needs are currently met using renewable resources.

Biomass is one of the most consistently and widely used renewable energy sources today. Biomass includes wood, charcoal (wood that has been baked to remove water and impurities), and animal waste products. Although biomass is renewable, as we discussed in Chapter 7, it is only renewable when it is used at a pace that allows time for replacement.

One interesting fuel that's recently been developed is called gasohol. **Gasohol** is a gasoline extender made from a mixture of 90 percent gasoline and 10 percent ethanol, which is often obtained by fermenting agricultural crops or crop wastes. So, as you can see, it is partly derived from organic substrate. Gasohol has higher octane than gasoline and burns more slowly, coolly, and completely, thus resulting in reduced emissions of some pollutants. Despite those advantages, it also vaporizes more readily than gasoline, and has the potential to aggravate ozone pollution in warm weather. Ethanol (and, similarly, methanol) from organic sources are surface carbon and therefore add no net carbon to the atmosphere. Additionally, because ethanol is carbon-based, carbon dioxide is released during its combustion. Ethanol-based gasohol is also expensive and energy-intensive to produce—one bushel of corn produces only two and a half gallons of ethanol.

This is clearly an inefficient way to use biomass. However, researchers are developing new and better ways to use biomass as a fuel; for example, a new product called **biodiesel** is made largely from waste vegetable oils.

Hydroelectric Energy

As its name suggests, **hydroelectric power** is generated by the force of water. Specifically, the electricity is generated as moving water turns a turbine. One advantage of hydroelectric power is that its production releases no pollutants. However, hydroelectric power does produce thermal pollution; in addition, it requires that rivers are dammed, and this can change the rates at which rivers flow and also lead to the destruction of habitats. On the other hand, as water is held behind dams, new habitats, in the form of wetlands, are created.

There are other problems associated with dams that you need to know about. One of them is silting. As water sits behind the dam, the normal sediments it carries have time to sink to the bottom. This puts additional weight on the structure and means that dams have to be built strong enough to hold back the many tons of sediment. This also means that the sediment is not passed farther down the river. The sediment that used to fertilize the flood plains of the river is now trapped behind the dam. In addition, the reservoir usually has a greater surface area than the preceding lake or river; this increased surface area can actually result in a higher rate of evaporation and water loss than before.

Another problem is that fish that spawn in the normally silty river no longer have a place to do so. Salmon and other anadromous fish breed in the streams where they hatched from eggs. Dams prevent the salmon from returning to their hatching streams. While **fish ladders** do let some fish return upriver, the number of fish that get through is so limited that the populations will still decline.

One more important note about hydroelectric power: The development of this as an alternative, renewable energy source is limited—simply because there is a limited number of rivers of sufficient flow and drop in the world that can be used for these purposes.

Solar Energy

While it is important to remember that we already obtain through the sun the energy we need to live—producers capture the sun's energy and convert it into chemical energy—solar energy also has the potential to supply many of our external energy needs as well.

The use of solar energy actually dates back to Roman times; the Romans developed window glass, which allowed sunlight to come in, and trapped solar heat indoors. Another interesting historical note is that the Swiss scientist Horace De Saussure built a solar reflector in 1767 that could heat water and cook food.

Passive solar energy collection is the use of building materials, building placement, and design to passively collect solar energy (such as through windows) that can be used to keep a building warm or cool. On the other hand, **active collection** is the use of devices, such as solar panels, that collect, focus, transport, or store solar energy. Solar panels absorb solar energy and pass the energy on to tubes in which water is circulating; this heated water can be stored for later use. Direct collection of solar energy via **photovoltaic cells** (PV cells) produces electricity, which is then stored in batteries. When sunlight hits the PV cells, electrons are energized and can flow freely, producing an electric current. After this, one of two things can happen. If the home or building is connected to a regional electric grid, the energy produced is fed into the grid; this results in the electricity meter on the building actually spinning backward! Homeowners who have installed solar panels receive a credit against further charges from their electricity providers when the energy that they've fed into the grid exceeds the amount of energy the household uses. If the home or building is not connected to the local electric grid, the energy stored can be stored in batteries to be utilized later.

While the use of solar energy produces no air pollutants, the production of photo-voltaic cells does require the use of fossil fuels. The advantages to solar panels are that photovoltaic cells use no moving parts, require little maintenance, and are silent. However, not every location receives enough sunlight to make solar panels worthwhile. Also, the initial financial outlay for solar power is significant, although money is saved when the home is disconnected to the regional grid. Additionally, some states (such as New Jersey) give homeowners financial assistance for the installation of solar systems in their homes. Eventually, new technology should significantly lower the cost of solar systems.

Wind Energy

People have been using wind to produce energy for centuries; as early as the seventeenth century windmills were so abundant in Schermerhorn (which is northwest of Amsterdam) that their turning paddles could be heard as far as 20 miles away! Windmills work in this way: Wind turns the blades, or paddles, of the windmill and this drives a shaft that's connected to several cogs. The cogs then turn wheels that can perform mechanical work, such as grinding grain or pumping water. Although the Dutch windmill is a picturesque symbol of wind power, the modern wind **turbine** looks more like an airplane propeller. The wind that blows into the wind turbine spins the blades, and this, in turn, causes the machinery inside the base of the windmill to rotate. The base of the windmill is called the **nacelle,** and it houses a gearbox and generator as well as machinery that controls the turbine. Wind turbines can be designed to utilize the energy from wind at all speeds, or to function only when the wind is at a certain velocity.

Wind energy is the fastest growing alternative energy source, and modern wind turbines are usually placed in groups called **wind farms** or parks. In the United States, the largest of these wind farms is located in Altamont Pass, California; this farm has several thousand wind turbines. Wind-generated power has been increasing at a rate of more than 30 percent per year and is projected to supply 3 percent of the world's energy needs by 2010 and perhaps up to 10 percent in Europe. Although in the United States wind farms are predominately in California and Texas at this time, many locations have enough prevailing winds to make production of electricity from wind power feasible. Wind farms can also be located offshore, in the ocean; and although they're currently only located near to shore, in the future they may be placed on floating docks in deep water.

At this time, wind power is more costly than using fossil fuels because of the initial outlay of capital that must be invested in order to build the windmills; windmills are also considered by many to be annoyingly loud and unattractive. However, perhaps the biggest problem with this type of renewable energy source is that alternate energy sources must be in place for times when there is no wind. In the 1990s one other public concern about the use of wind turbines was that birds would be cut up and killed by the blades, but now we know that as long as wind farms are not located in the middle of migration routes, only one or two birds per turbine per year are killed—and this is far fewer than the number of birds killed by other types of towers. Finally, one tremendous advantage of using wind energy is that it produces no harmful emissions. Let's move on and talk about another type of renewable energy source—geothermal energy.

Geothermal Energy

Geothermal energy is a form of energy that's obtained from within the earth; it's energy that's produced by harnessing the earth's internal heat. The interior of the earth is still warm due to radioactive decay. Therefore, geothermal energy indirectly gains its energy from nuclear power. The greatly elevated temperatures within the earth result in a buildup of pressure; some of this heat escapes through fissures and cracks to the surface. Some common examples of these fissures and cracks that you may have heard of are geysers, hydrothermal vents, and hot springs.

More specifically, in the process of geothermal energy production, the naturally heated water and steam from the earth's interior turn turbines, and this creates electricity. Although surface water from geysers could be used, wells are typically drilled down into the earth as far as thousands of meters to water that is 300–700 degrees Fahrenheit and then brought to the surface and converted to steam, which powers a turbine. Geothermal energy can also be used directly; in this process, the heated water is piped directly though buildings to heat them—this is a common method for heating homes in Iceland. In a sense, geothermal energy is renewable; however, if the groundwater is used at a faster rate than it is replaced, then this energy source is limited.

The use of geothermal energy is also limited because only a few areas have geothermal sources to tap. Another problem with this renewable energy source is that the salts that are dissolved in the water corrode machinery parts; additionally, some gases (such as methane, carbon dioxide, hydrogen sulfide, and ammonia) that are trapped in the water may be released as the water is utilized.

Still Other Sources of Energy

There are two other less widespread renewable energy sources: energy that can be harnessed from tidal movement in the ocean, and hydrogen cells. You should be somewhat familiar with both of these energy sources for the exam.

Ocean Tides

The tidal movements of ocean water can be tapped and used as a source of energy. To harvest tidal energy, dams are erected across outlets of tidal basins. Incoming tides are sluiced through the dam, and the outgoing tides pass through the dam, turning turbines and generating electricity. Recently, ocean dams have been developed that allow energy to be harnessed from both the outgoing and incoming tides.

At this time, there is a tidal power plant installed in the East River in New York, NY. It harnesses enough power to power 10,000 homes. There are many different designs of ocean tide power plants in the idea stages. One of these involves having the wave push into a chamber of air; the compressed air is then forced through a small hole at the turbine, and turns the turbine as it is released. An experimental prototype of this design has been installed off the coast of Scotland and is nicknamed the LIMPET (Land-Installed-Marine-Powered Energy Transformer).

Hydrogen Cells

Hydrogen is obtained from fossil fuels by a process called reforming. Hydrogen is very difficult to store and not very energy dense, but Hydrogen fuel cells are considered by many to be the best, cleanest, and safest fuel source. Free hydrogen is not found on Earth, but it can be released through the process of electrolysis, in which hydrogen atoms are stripped from water, leaving the oxygen atom. Hydrogen can also be obtained from organic molecules, but the use of organic sources can release pollutants—as can the process of electrolysis if a fossil fuel, such as natural gas or coal, is used to drive the process. However, once the free hydrogen is released, it can be stored and then used to generate electricity through the reverse reaction of electrolysis.

One of the major benefits of the use of hydrogen fuel cells is that the only waste from the fuel cell is steam—water vapor. This technology has been used for decades in spacecrafts, but the high cost of the fuel cell and lack of hydrogen fuel stations has limited the technology to just a few test programs. The United States Department of Energy estimates that hydrogen fuel cells large enough to power light trucks and cars in the United States will require the production of 150 megatons of hydrogen per year (in 2004, only nine megatons of hydrogen were produced). Co-generation is also an energy source in which the waste heat from electricity generation is used in another industrial or residential process thereby increasing the efficiency of the use of that fuel. It is possible to use the combustion of a fuel for more than one purpose simultaneously. Often a generator is driven by burning natural gas near the turbine. The gases are still very hot after they pass by the turbine. This formerly "waste" heat can then be used to transform water to steam to heat buildings or to run a chiller to cool buildings through a deep application of thermodynamic principles. Either way, the efficiency of the system is raised considerably by the dual use of the fuel combusted.

In order for hydrogen to become a truly viable option as a renewable energy source, an inexpensive and efficient way to produce hydrogen from nonfossil fuel sources must be developed. One of the most promising techniques for this involves the use of photovoltaic cells to harvest sunlight and then power the splitting of the water molecule.

ENERGY CONSERVATION, A FINAL NOTE

When we discuss **energy conservation**, we are basically referring to the practice of reducing our use of fossil fuels and reducing the impact we have on the environment as we produce and use energy.

One important form of energy conservation is the use of alternative fuel cars. They are gaining in popularity and acceptance. Hybrid vehicles are built with two motors: one electric and one gasoline-powered. The electric motor powers the car from 0 to about 35 miles per hour. Above 35 mph, the gasoline engine starts and powers the car. At highway speeds, both the electric and gas motors operate. The cars are designed so that when the brakes are applied, some of the energy is transferred from the brakes to recharge the electric motor's battery. Not only do these cars have good gas mileage, but they also produce far less CO_2 pollution. Although not as common as hybrids, several carmakers make models that use propane or natural gas as fuels. These generate only CO_2 and water as emissions, and they get good gas mileage. A problem is the lack of refueling stations, although devices are available that allow refueling from home. Cars can also be retrofitted with natural gas fuel tanks, so the driver can choose between gasoline or methane fuel.

Another type of alternative fuel is used cooking oils. The oils used in deep-fat fryers can be filtered and then burned in diesel-fueled cars, trucks, and buses. After starting the engine on pure diesel fuel, the driver switches to the **biofuel** to drive. At the end of the trip, the driver runs on pure diesel fuel again for a few minutes before shutting off the engine.

It has been argued that finding new fossil fuel sources would serve the same purpose as would reducing our current use of fossil fuels. However, this is not true—this statement does not take into consideration the fact that our use of fossil fuels has numerous negative effects on the environment. Additionally, in the long term it will not help us much to conserve fossil fuel resources—simply because these are not renewable energy sources—so they will eventually be depleted. Therefore, if we are to have dependable long-term renewable sources of energy, we must continue to develop, implement, and improve upon current technology and methods.

On the legislative front, the United States has adapted the **CAFE, or Corporate Average Fuel Economy**, standards. These standards set mile per gallon standards for a fleet of cars. The goal of these standards is to reduce energy consumption by increasing the fuel economy of cars and light trucks. Review the CAFE standards described on page 190.

Finally, do not forget the role of mass transit in reducing pollution. Buses and trains can move many more people than cars. When the amount of pollution made by the vehicle is divided by all the passengers it is carrying, the bus or train generates far less pollution per person than a car.

In the next chapter, we will further discuss pollution—the effects it has on Earth and its inhabitants, and the types of wastes that currently exist, and how we can manage them.

Before you move on, study the terms and try the questions in the drill on the following pages. Good luck!

KEY TERMS

Use some of your renewable brain energy to study these terms.

Energy
potential energy
kinetic energy
radiant energy
conduction
convection
energy quality
net energy yield
energy units
power units
first and second laws of
        thermodynamics

Non Renewable Energy
fossil fuels
seam
exploratory well
proven reserve
crude oil
petroleum
anthracite
bituminous coal
subbituminous coal
lignite
underground mining
fly ash
boiler residue
strip mining
overburden
scrubbers
acid mine drainage

Nuclear Energy
fission
nuclear fusion
breeder reactor
half-life
fuel rod
boiling water reactor
pressurized water reactor
NIMBY

Renewable Energy
gasohol
biodiesel
hydroelectric power
fish ladder
passive collection
active collection
photovoltaic cells
nacelle
turbine
wind farm
geothermal energy
hydrogen cell
ocean tides
CAFE standards

Chapter 8 Drill

Directions: Each of the questions or incomplete statements below is followed by five suggested answers or completions. Select the one that is best in each case. For answers and explanations, see Chapter 13.

1. A fuel's net energy yield is correctly defined as

    (A) how much of this fuel is left in the world
    (B) how much time it takes to extract and transport
    (C) a comparison between the amount of pollution the fuel generates and the amount of useful energy produced
    (D) a comparison between the costs of mining, processing, and transporting a fuel and the amount of useful energy the fuel generates
    (E) a comparison between the amount of fuel in reserve and the speed at which the fuel is being removed

2. A regular light bulb has an efficiency rating of 3 percent. For every 1.00 joule of energy that bulb uses, the amount of useful energy produced is

    (A) 1.03 joules of light
    (B) 1.03 joules of heat
    (C) 0.97 joules of light
    (D) 0.03 joules of light
    (E) 0.03 joules of heat

3. Methane gas and ethanol are two examples of biogases that are produced in which of the following processes?

    (A) The distillation of oil
    (B) The pressurization of natural gas
    (C) The anaerobic digestion of biomass
    (D) The catalytic reaction of coal and limestone
    (E) The breakdown of water by electricity

4. Hybrid car engines have which of the following types of motors?

    (A) Gasoline powered only
    (B) Natural gas powered only
    (C) Electric powered only
    (D) Gasoline and natural gas powered engines
    (E) Gasoline and electric powered engines

5. All of the following are ways to increase energy efficiency EXCEPT

    (A) using low volume shower spray heads
    (B) insulating your home thoroughly
    (C) switching incandescent light bulbs to fluorescent bulbs
    (D) leaving room lights on
    (E) increasing fuel efficiency of vehicles

6. A typical coal-burning power plant uses 4,500 tons of coal per day. Each pound of coal produces 5,000 BTUs of electrical energy. How many BTUs are produced each day from this plant?

    (A) 4.5×10^{10}
    (B) 0.45×10^{10}
    (C) 11.5×10^3
    (D) 4.5×10^8
    (E) 0.25×10^6

7. Which of the following produces the least amount of carbon dioxide while generating electricity?

    (A) Oil
    (B) Coal
    (C) Wind turbines
    (D) Wood
    (E) Diesel fuel

8. How much energy, in kWh, is used by a 100-watt computer running for 5 hours?

    (A) 500 kWh
    (B) 200 kWh
    (C) 100 kWh
    (D) 50 kWh
    (E) 0.5 kWh

9. Photovoltaic cells produce electricity by

    (A) a system of mirrors that focuses sunlight onto a heat collection device
    (B) using the sun's energy to create a flow of electrons in a material such as silicon
    (C) breaking down organic molecules and releasing energy
    (D) warming air, which spins a turbine
    (E) acting as a catalyst to burn oil cleanly

10. A sample of radioactive material has a half-life of 20 years. It has an activity of 2 curies. How many years does it take for the material to have an activity level of 0.25 curies?

    (A) 20 years
    (B) 40 years
    (C) 60 years
    (D) 80 years
    (E) 100 years

11. The term vampire appliances correctly refers to appliances that

(A) generate more power than they consume
(B) consume electricity even when they are not operating
(C) are EnergyStar rated
(D) are programmed to turn themselves off at midnight each night
(E) have more than one transformer

12. All non-renewable resource power plants use heat to

(A) make hot air that generates power
(B) create powerful magnetic fields that make electricity
(C) create powerful water jets that spin turbines
(D) produce steam to turn electric generators
(E) split water into hydrogen and oxygen that is then burned to make electricity

Free-Response Question

1. Nuclear power plants have been described as being part of the solution to the problem of the United States' dependency on foreign energy. Currently, some 20 percent of the electricity produced in the United States is generated by nuclear power.

    (a) Describe the key parts of a nuclear power plant. Describe the roles of the following: core, fuel rods, coolant, and heat exchanger.
    (b) Describe two practical methods of dealing with the long-term storage of the highly radioactive wastes produced by a power plant.
    (c) Describe one positive impact that a nuclear power plant might have on air pollution.
    (d) Opponents of nuclear power plants point out the problems caused by thermal pollution of nearby rivers. Describe how the thermal pollution occurs and one method to reduce this problem.

REFLECT ACTIVITY

Respond to the following questions:

- For which content topics discussed in this chapter do you feel you have achieved sufficient mastery to answer multiple-choice questions correctly?

- For which content topics discussed in this chapter do you feel you have achieved sufficient mastery to identify hot-button topics and successfully articulate them in free-response questions?

- For which content topics discussed in this chapter do you feel you need more work before you can answer multiple-choice questions correctly?

- For which content topics discussed in this chapter do you feel you need more work before you can successfully identify and articulate them in free-response questions?

- What parts of this chapter are you going to re-review?

- Will you seek further help, outside of this book (such as a teacher, tutor, or AP Central), on any of the content in this chapter—and, if so, on what content?

Chapter 9
Pollution

According to the College Board, about 25–30 percent of the questions you'll see on this exam will be on pollution. That means that there will be more questions about topics that fall under the broad category of pollution than there will be questions about any of the other major topics. This just goes to show the environmental importance of how we create waste and what we do with it.

We'll begin the chapter with a discussion of what it means for something to be toxic. We'll move on to discuss toxins in air pollution, and then review the major aspects of thermal pollution, water pollution, and the problems that arise as a result of solid waste. As we go through each type of pollution, we'll also discuss the impact of pollution on the environment and human health, as well as some economic impacts. Let's begin!

TOXICITY AND HEALTH

A **toxin** is any substance that is inhaled, ingested, or absorbed at sufficient dosages that it damages a living organism, and the **toxicity** of a toxin is the degree to which it is biologically harmful. Almost any substance that is inhaled, ingested, or absorbed by a living organism can be harmful when it is present in large enough quantities—even water!

Substances are usually tested for toxicity using a **dose-response analysis.** In a dose-response analysis, organisms are exposed to a toxin at different concentrations, and the dosage that causes the death of the organism is recorded. The information from a set of organisms is graphed, and the resulting curve is referred to as a **dose-response curve**. The dosage of toxin it takes to kill 50 percent of the test animals is termed LD_{50}, and this value can be determined from the graph. A high LD_{50} indicates that a substance has a low toxicity; a low one indicates high toxicity. A **poison** is any substance that has an LD_{50} of 50 mg or less per kg of body weight.

If just the negative health effects are plotted, instead of the level of the toxin at which death occurs, the resulting graph indicates the dosage that causes a change in the state of health. In this case, the ED_{50} is the point at which 50 percent of the test organisms show a negative effect from the toxin. The dosage at which a negative effect occurs is referred to as the **threshold dose**. Two more terms you should know for the test are acute effect and **chronic effect.** An **acute effect** is an effect caused by a short exposure to a high level of toxin; a snakebite, for example, causes an acute effect. A **chronic effect** is what results from long-term exposure to low levels of toxin; an example of this would be long-term exposure to lead paint in a house.

As they say, dose makes the poison. In order for a substance to be harmful, all of the following must be considered:

- Dosage amount over a period of time

- Number of times of exposure

- Size and/or age of the organism that is exposed

- Ability of the body to detoxify that substance

- Organism's sensitivity to that substance (due, for example, to genetic predisposition, or previous exposure)

- Synergistic effect (more than one substance combines to cause a toxic effect that's greater than any one component)

An **infection** is the result of a pathogen invading our body, and **disease** occurs when the infection causes a change in the state of health. For example, HIV, the virus that causes the disease AIDS, infects the body and typically has a long residence time. When it causes a change in a person's state of health, it has morphed into a disease called AIDS.

There are five main categories of pathogens.

- Viruses (and other subcellular infectious particles, such as prions)

- Bacteria

- Fungi

- Protozoa

- Parasitic worms

Pathogens can attack directly or via a carrier organism (called a **vector**). One example of a pathogen that relies on a vector is the bacteria that causes Rocky Mountain spotted fever; it lives in the bodies of ticks, and when ticks bite humans, the ticks inject the bacteria and cause the disease.

As you're probably well aware, other things besides pathogens can make people ill, including environmental factors such as tobacco smoke, UV radiation, or asbestos. Also, although you may be exposed to a toxin or an infectious agent and not experience a change in the state of your health, someone else who's exposed to the toxic agent or pathogen could become very ill.

The degree of likelihood that a person will become ill after exposure to a toxin or pathogen is called **risk.** Many environmental, medical, and public health decisions are based on potential risk. Calculating risk is referred to as **risk assessment,** and **risk management** means using strategies to reduce the amount of risk. The U.S. Department of Public Health and Public Services is an organization that makes use of risk assessment and management; for example, they decide who can receive the flu shot each year. If the risk of getting the flu is high for a particular year, most of the population is encouraged to get the shot; however, if the risk seems small or the predicted flu strains are mild, only older people and the immunocompromised are advised to get the flu shot.

AIR POLLUTION

Substances that are considered contributors to air pollution have two sources; they can be natural releases from the environment or they can be created by humans. The effects of air pollution on humans can range in severity from lethal to simply aggravating. Some natural pollutants include pollen, dust particles, mold spores, forest fires, and volcanic gases. One of the more recently described air pollutants from nature is produced by dinoflagellates, which, you might recall from Chapter 4, are the organisms that cause red tide. The toxins that are produced by these algae are caught in sea spray, in which they can be aerosolized and inhaled by humans, causing respiratory distress.

Although you may think that human-caused pollution is a relatively new phenomenon, people have added pollutants to the air throughout the history of humankind. Early man's fire created pollutants, and the Roman's smelting of lead resulted in air pollution that drifted thousands of miles from the source—and has even been discovered trapped in the ice of Greenland! It is true, however, that the large-scale production of pollutants began with the Industrial Revolution; and this is especially true of air pollution. The beginning of the Industrial Revolution marked the entrance of pollutants from fossil fuel into the atmosphere, for example, which has been environmentally disastrous.

Let's go over some terms used to describe pollution before we get into more specific details. **Primary pollutants** are those that are released directly into the lower atmosphere (remember the troposphere?) and are toxic; one example of a primary pollutant is carbon monoxide (CO). **Secondary pollutants** are those that are formed by the combination of primary pollutants in the atmosphere; an example of a secondary pollutant is acid rain. Acid rain is produced from the combination of **sulfur oxides** (such as SO_2 and SO_3) and water vapor.

Pollutants can be released by **stationary sources,** such as factories or power plants, or they can be released by **moving sources,** like cars. **Point source pollution** describes a specific location from which pollution is released; an example of a point source location might be a factory or a site where wood is being burned. Pollution that does not have a specific point of release—for example, a combination of many sources, such as a number of cows releasing methane gas within a few square miles—is known as **non-point source pollution.**

The Major Culprits

The Environmental Protection Agency has determined that there are six pollutants (familiarly referred to as the dirty half dozen) that do the most harm to human health and welfare; the Environmental Protection Agency (EPA) refers to them as **criteria pollutants.**

The Six Criteria Pollutants

- carbon monoxide, CO

- lead, Pb

- ozone, O_3

- nitrogen dioxide, NO_2

- sulfur dioxide, SO_2

- particulates

In general, gases in the atmosphere are measured in units of parts per million, or ppm, when they are in relative abundance; when they are present in trace (very small) amounts, they are measured in parts per billion (ppb). For example, if in a certain geographic area, the carbon dioxide content of the air is 10 ppm, this would mean that there are ten molecules of CO_2 per one million molecules of air. These pollutants are monitored with the National Ambient Air Quality Standards (NAAQS). These standards were established by the Environmental Protection Agency (EPA) to protect human health.

Let's go back through the list above. Carbon monoxide (CO) is an odorless, colorless gas that's typically released as a by-product of incompletely burned organic material, such as fossil fuels. CO is hazardous to human health because it binds irreversibly to hemoglobin in the blood. Hemoglobin is the molecule that is responsible for transporting oxygen around the body from the lungs. Hemoglobin has a higher affinity for CO than it does for oxygen, which means that in the presence of both CO and O_2, CO will bind more readily than O_2. In our normal oxygen-rich environments this competition is not a problem, but in areas where CO is present in large concentrations, it can be deadly. More than 60 percent of the CO released into the atmosphere comes from vehicles that burn fossil fuels.

Lead is an air pollutant that, as you now know, has been around since the time of the Roman smelters. It is generally released into the atmosphere as a particulate (a very small solid particle that can be suspended in the air), but then settles on land and water, where it is incorporated into the food chain and is part of biomagnification. If it enters the human body, it can cause numerous nervous system disorders, including mental retardation in children. At one time, lead entered the atmosphere primarily as a result of the burning of leaded gasoline. However, lead gas has been phased out, and now the primary source of lead is industrial smelting. Incidentally, the "lead" in your pencils is not the element lead; in fact, it's the mineral graphite. The graphite in pencils received the name "lead" because of its lead-like color when it's transferred to paper.

We began our discussion of ozone in Chapter 4 and have mentioned it several times since. Notice that the ozone the EPA calls one of the dirty half dozen is specifically—and only—the ozone that's formed as a result of human activity. All ozone is O_3 and is the same chemically. However, **stratospheric ozone** (absorbs UV light from the Sun and therefore protects life on our planet) is functionally very different from **tropospheric ozone** (powerful respiratory irritant and precursor to secondary air pollutants). Up high, ozone helps us; down low, it hurts us. O_3 is a secondary pollutant; it is formed in the troposphere as a result of the interaction of **nitrogen oxides**, heat, sunlight, and volatile organic compounds (VOCs—more on this later). Tropospheric ozone is a major component of what we think of as smog (more on this later).

The next major culprit on the list, nitrogen dioxide (NO_2), is one in a family of nitrogen and oxygen gases. NO_2 and the other nitrogen oxides are formed when atmospheric nitrogen and oxygen react as a result of exposure to high temperatures; this type of reaction occurs in combustion engines, for example. In fact, more than half of the nitrogen oxides in the atmosphere are released as a result of combustion engines. Other sources of nitrogen oxides are utilities and industrial combustion. Nitrogen dioxide is also commonly found as a secondary pollutant, and is a component of smog and acid precipitation.

Sulfur dioxide (SO_2) is a colorless gas with a penetrating and suffocating odor. It is a powerful respiratory irritant, and is typically released into the air through the combustion of coal. As we mentioned in Chapter 8, the use of scrubbers in coal-burning plants has helped reduce the amount of SO_2 released into the atmosphere. However, there are other sources of sulfur dioxide, including the processes of metal smelting, paper pulping, and the burning of fossil fuels. Sulfur dioxide can also be a component of indoor pollutant as a result of gas heaters, improperly vented gas ranges, and tobacco smoke. In the atmosphere, SO_2 reacts with water vapor to form acid precipitation. Here's one last note to help you with the test: Both nitrogen and sulfur can combine with oxygen to make several different molecules. Rather than a list of all the possible molecules, you might see the terms NO_x and SO_x. These terms (O_x) mean that there are several sulfur- and nitrogen-containing compounds mixed together.

Particulate matter is the last on the EPA's list of the dirty half dozen. Like lead, it is not a gas, but exists in the form of small particles of solid or liquid material. These particles are light enough to be carried on air currents, and when humans breathe them in, the particles act as irritants. Examples include soot (black carbon) and sulfate aerosols.

There have been significant decreases in the atmospheric content of both lead and carbon monoxide since the 1970s, mostly because of the phasing out of lead gasoline and the introduction of car engines that burn more cleanly. However, there are other air pollutants that are a growing concern to environmentalists, including the volatile organic compounds (VOCs), which are released as a result of various industrial processes including dry cleaning, the use of industrial solvents, and the use of propane. VOCs can react in the atmosphere with other gases to form O_3 and are a major contributor to smog in urban areas. Now, what exactly is smog?

Smog

As you might be aware, the setting for many of the Sherlock Holmes mysteries was the foggy, smoggy city of London. The smog that covered London throughout the nineteenth century, and well into the middle of the twentieth century, was **industrial smog**—also known as **gray smog**. As deadly as any of Holmes's adversaries in Sir Arthur Conan Doyle's stories, gray smog killed more than 2,000 people in a prolonged smog incident in 1911. However, the worst pollution-related incident in London occurred in 1952 and led to the death of about 10,000 city dwellers from pneumonia, tuberculosis, heart failure, and bronchitis. It was this disaster, resulting from the burning of large amounts of low-quality coal to heat homes and combat a cold fog, that prompted the Clean Air Act of 1952 in England.

Industrial smog is formed from pollutants that are typically associated with the burning of oil or coal. When **CO and CO_2** are released in the process of combustion, they combine with particulate matter in the atmosphere and produce smog. The production of smog can also be aided by weather conditions—air inversions, for example, which trap the pollutants; or fog, which holds the pollutants. As we mentioned above, sulfur dioxide may be another component in gray smog, combining with water vapor to form sulfuric acid that is suspended in the cloud of smog.

Photochemical smog—also known as **brown smog**—a different type of smog, is usually formed on hot, sunny days in urban areas. In photochemical smog, NO_x compounds, VOCs, and ozone all combine to form smog with a brownish hue. The intensity of sunlight on these days also promotes the formation of ozone from the combination of NO_x compounds. Los Angeles, California and Athens, Greece are two cities that are particularly susceptible to photochemical smog. Athens has enacted mandates that have already reduced the number of cars driven each day in the city and improved the quality of the air. For example, in Athens, by law, cars with even numbered license plates can only be driven on even-numbered days—and cars with odd-numbered license plates can only be driven on odd-numbered days!

Ozone Depletion

While harmful in the troposphere, as you know, ozone in the stratosphere provides us with a much-needed defense against ultraviolet radiation. The ozone layer is responsible for blocking about 95 percent of the sun's ultraviolet radiation (UV), thus protecting surface-dwelling organisms from UV damage. Ozone is naturally created by the interaction of sunlight and atmospheric oxygen. The simplified reaction is

$$O_2 + UV \text{ (sunlight)} \rightarrow O + O$$
$$O + O_2 \rightarrow O_3$$

As early as the mid-1950s, a thinning of the ozone layer above the Antarctic was observed. In the 1970s, atmospheric scientists hypothesized, and later proved, that declining stratospheric ozone levels were due a group of man-made chemicals known as **chlorofluorocarbons (CFCs)**. Invented in the 1930s, CFCs and many other related compounds (e.g., halons and hydrochlorofluorocarbons) were used in items such as propellants, fire extinguishers, and cans of hairspray.

Once released, CFCs migrate to the stratosphere through atmospheric mixing (they are very stable, which allows them to survive through the rise). In the upper stratosphere, intense UV radiation breaks the CFC molecules apart and releases chlorine atoms that form chlorine monoxide (ClO) while converting O_3 to O_2. Let's take a look at that reaction.

$$Cl + O_3 \rightarrow ClO + O_2$$

During the winter months, chlorine monoxide is concentrated on ice crystals that form in and around the Antarctic polar vortex. In early spring, the returning warmth of the sun frees the chlorine from the chlorine monoxide where it destroys more ozone. The reaction that frees the chlorine from chlorine monoxide is

$$ClO + O \rightarrow Cl + O_2$$

Ozone loss is greatest in the spring as the chlorine breaks down ozone into O_2. Remember that chlorine acts as a catalyst; it is not changed by its reaction with ozone and it can help break down another O_3 molecule immediately. As the air continues to warm, the natural production of ozone *increases* as more sunlight catalyzes the combination of oxygen back into ozone. This occurs in January and February (Antarctica's summer).

The Antarctic continent is the area exposed to the greatest amount of UV radiation, but prevailing winds can carry the ozone-depleted air to South America, Australia, and southern Africa. In 2006, the area of ozone thinness was over 26 million square kilometers. Reduced levels of ozone have been documented over the Arctic and even over some midlatitude regions.

The loss of ozone has serious implications for the earth's ecosystems, as well as for human health. The increased number of UV rays that reach Earth through the thin ozone layer can kill phytoplankton and other primary producers. The decrease in primary productivity of both marine and terrestrial ecosystems lowers the amount of available fish and crops. Human health issues from increased exposure to UV rays include eye cataracts, skin cancers, and the weakening of our immune systems.

Now for the good news: There are several methods to manage the amounts of CFCs. In 1987 the Montreal Protocol was signed by more than 146 nations. The protocol calls for the worldwide end of CFC production. The United States stopped production in 1995. Since the institution of the Montreal Protocol, the release of ozone-depleting chemicals has been reduced by 95 percent. There are, however, many nations that still rely on CFCs, though work is being conducted to develop safe and effective substitutes.

Acid Rain

Acid precipitation—in the form of acid rain, acid hail, acid snow, etc.—occurs as a result of pollution in the atmosphere; primarily SO_2 and nitrogen oxides. These gases combine with water to form acids (typically nitric acid and sulfuric acid) that are deposited on the earth through precipitation. Because this acid is highly diluted, acid precipitation isn't acidic enough to burn the skin upon contact, but it does have a significant, measurable effect on humans and the environment. How acidic is acid rain? Well, rain usually has a pH of about 5.6, but acid rain can have a pH as low as 2.3. The pH of rain is not 7.0 because of the carbonic acid in rainwater. Also, it's important for you to know that the pH scale is logarithmic, meaning that each whole pH value below 7 is ten times more acidic than the next higher value.

Acid precipitation can be a chronic and significant problem for large urban areas with many vehicles, and areas that are downwind of coal burning plants. While **dry acid particle deposition** occurs two to three days after emission into the atmosphere, **wet deposition** is usually delayed for four to fourteen days after emission; therefore it can travel in air currents to locations that are many miles downwind of the emission source.

Some areas, like those with already acidic soils that were derived from granite, are particularly vulnerable to acid precipitation. Other areas that are particularly vulnerable to acid precipitation are those where the soil has been leached of its natural calcium content. This is because calcium acts as a natural buffer and would temper the effects of acid precipitation. Acid rain is a significant cause of sink holes in Florida due to acid dissolving the limestone rock that much of Florida is made of.

In some areas of the world, progress has been made toward controlling acid precipitation. The 1990 amendment to the Clean Air Act (CAA) has led to significant reductions in the amounts of SO_2 and NO_x that are emitted from industrial plants. Despite **National Ambient Air Quality Standards**

Acid precipitation is responsible for the following effects:

- Leaching of some minerals from soil (which alters soil chemistry)

- Creating a buildup of sulfur and nitrogen ions in soil

- Increasing the aluminum concentration in soil to levels that are toxic for plants

- Leaching calcium ions from the needles of conifers

- Elevating the aluminum concentration in lakes to levels that are toxic to fish

- Lowering the pH of streams, rivers, ponds, and lakes, which may lead to fish kills

- Causing human respiratory irritation

- Damaging all types of rocks, including statues, monuments, and buildings (**standards established by the United States Environmental Protection Agency**), there is still considerable damage being done to soils and lakes in many areas, and these ecosystems will not be able to continue to tolerate significant lowering of their pH.

Motor Vehicles and Air Pollution

Today, all new vehicles sold in the United States must meet the EPA standards (in California, they must meet certain standards set by the state). Due to the Clean Air Act (the CAA) and its amendment (the CAAA), new cars (those produced after the year 1999) emit 75 percent fewer pollutants than cars made before 1970. The most significant device in controlling emissions in cars is the **catalytic converter**. This platinum-coated device oxidizes most of the VOCs and some of the CO that would otherwise be emitted in exhaust, converting them to CO_2. Newer models of catalytic converters also reduce nitrogen oxides, but not very successfully.

As mentioned in Chapter 8, in the Energy Policy and Conservation Act of 1975, the Department of Transportation (DOTS) was given the authority to set what's called **Corporate Average Fuel Economy (CAFE)** for motor vehicles. CAFE was intended to reduce both fuel consumption and emissions (not surprisingly, because burning less gas creates less air pollution). The standard today requires that vehicles have a fuel efficiency average of 27.5 miles per gallon, but larger vehicles such as pick-up trucks, SUVs, and minivans have a lower standard of just 22.7 mpg. Under new Federal Tier 2 standards, which went into effect in 2007, for the first time, light trucks will be held to the same standards as passenger cars.

Tier 2 standards also limit nitrous oxide (NO) emission to 0.07 grams per mile, which represents a reduction of 90 percent for passenger cars—and even more for light trucks. There is also a target reduction for sulfur emissions from gasoline; the new standards will decrease acceptable emissions from 300 ppm to 30 ppm.

All of these new standards will most likely result in higher purchase prices for vehicles, and they have certainly caused an outcry from auto manufacturer and oil refineries. However, the new standards are expected to reduce air pollutants by two million tons per year.

Vehicles of the Future

In 1990, the state of California passed a No-Pollution Vehicle Law mandating that, by 2003, 10 percent of the cars sold in the state would be pollution free. That law was later rescinded because of problems with the development of the zero pollution electric car, which looked promising at the time the bill was passed. In the past, electric cars were not widely adopted because of a limited traveling range, they weighed much less than their gasoline burning counterparts, and they lacked amenities (like air conditioning). However, a renewed interest as well as new battery technology, which increases energy storage and reduces cost, has motivated car makers to produce a promising generation of fully electric cars. The Nissan

Leaf and Cheverolet Volt, both available for purchase, are examples of some zero pollution electric vehicles.

In addition to new electric cars, hybrid vehicles that run on a mixture of gas and electric power have been gaining popularity over the past few years. Government regulations, incentives, and public acceptance will probably determine how quickly the hybrid car moves into the mainstream vehicle market. One incentive that's been offered at a federal level is a full-dollar tax credit. The amount of the credit varies by car model. That incentive is scheduled to be reduced unless a new energy bill changes it. Some individual states also provide incentives to residents who purchase a hybrid vehicle.

It is highly unlikely that Congress will enact legislation that will provide real incentives for the purchase of hybrid vehicles or other alternatives that would reduce air pollution from vehicles. This is in part due to the fact that lobbying groups representing the oil companies and vehicle manufactures consistently lobby against these incentives. However, in the future, grassroots organizations that are backed by the voting public may influence legislation.

A hydrogen fuel cell vehicle would produce even less pollution than a hybrid vehicle, but don't expect to see them on the market very soon. Mass producing the cells is still not cheap enough to make the cars economically viable, as we mentioned in Chapter 8.

Indoor Air Pollution

The idea of air pollution that exists indoors, and the concept of the condition "sick building syndrome" is still relatively new, but it is now widely recognized that air pollutants are usually at a higher concentration indoors than outside. This makes sense if you consider that pollutants that exist outside can also move inside as doors and windows are opened. Once the pollutant is indoors, it remains trapped until air currents move it out the door or windows or though a ventilation system. Additionally, indoor spaces have certain pollutants that are unique to them. The World Health Organization (WHO) estimates that indoor air pollution is responsible for 1.6 million annual deaths worldwide (that's one death every 20 seconds!). According to the Environmental Protection Agency (the EPA), indoor air pollution is one of the five major environmental risks to human heath.

One reason that indoor air pollution has such a great impact on health is because humans spend a significant amount of time indoors. Especially in developed countries, people generally work and live in well-sealed buildings that have little air exchange. In developing countries, however, one of the worst indoor air pollutants is material that's used for fuel. Dung, wood, and crop waste are the primary fuels used by more than half the world's population in order to heat homes and cook food, and the particulate matter that results from burning these fuels can exceed acceptable levels by hundreds of times.

In developed countries, other pollutants play the biggest roles in the creation of indoor air pollution; the most abundant indoor pollutants are **volatile organic compounds (VOCs)**. VOCs are found in carpet, furniture, plastic, oils, paints, adhesives, pesticides, and cleaning fluids. Even dishwashers are responsible for the creation of VOCs, when chlorine detergent reacts with leftover foods. Another component of pollution in developed countries is CO; CO arises in indoor air as a result of gas leaks or poor gas combustion devices. CO detectors are available for homes, and can prevent CO poisoning.

Two of the most deadly and common indoor pollutants in developed countries are tobacco smoke and radon. Tobacco smoke affects not only the health of the smoker, but the health of those around the smoker as well. Secondhand smoke causes the same symptoms in nonsmokers who simply breathe in secondhand smoke. Secondhand smoke, which contains over 4,000 different chemicals, has been classified by the EPA as a Group A carcinogen (meaning that it causes cancer in humans). It's estimated that secondhand smoke causes 35,000–40,000 deaths per year from heart disease, and 3,000 deaths from lung cancer. In children younger than 18 months, it is responsible for 150,000–300,000 lower respiratory tract infections annually, and increases the number and severity of asthma attacks in about one million asthmatic children.

Radon is the second leading cause of lung cancer (after smoking) in the United States. Radon is a gas that's emitted by uranium as it undergoes radioactive decay. It seeps up through rocks and soil and enters buildings. It is not found everywhere, and must be tested for specifically. Homes that were built after 1990 have radon-resistant features.

The final indoor pollutants we'll review are actually living: Certain living organisms, such as tiny insects, fungi, and bacteria, are considered pollutants. Many people are allergic to mold spores, mites, and animal dander, but asthma attacks can also be triggered by these living pollutants. The water tanks for large air conditioning units are good places for certain types of bacteria to grow, and as air is distributed throughout the house, the bacteria are also distributed. Some bacteria can cause diseases; one example of this is *Legionella pneumophila*, which causes Legionnaires disease.

Sick Building Syndrome

Sick building syndrome (SBS) is a term that's used when the majority of a building's occupants experience certain symptoms that vary with the amount of time spent in the building, and for which no other cause can be identified. SBS is somewhat difficult to diagnose, and specific culprits are very difficult to identify. A condition is referred to as a **building-related illness** when the signs and symptoms can be attributed to a specific infectious organism that resides in the building. One example of a building-related illness is Legionnaires disease. Some symptoms of SBS include the following:

- irritation of the eyes, nose, and throat

- neurological symptoms, such as headaches and dizziness; reduction in the ability to concentrate; or memory loss

- skin irritation

- nausea or vomiting

- a change in odor or taste sensitivity

There are many ways in which people can reduce the amount of indoor pollutants that they're exposed to—for many people simply quitting smoking or encouraging roommates to quit would make a huge difference. Other precautions that people can take are to limit the amount of exposure they have to certain chemicals, such as pesticides or cleaning fluids. Perhaps the most important step to take is making sure that buildings are as well ventilated as possible.

CLIMATE CHANGE

Scientists use very sophisticated computer models and make several thousand meteorological observations each day to monitor the daily temperature of the earth's atmosphere. Over the last several years, their observations have shown that there has been a slow but steady rise in the earth's average temperature. The summers of 1998, 2005, and 2010 were the warmest on record. Other qualified scientists have carefully documented a decrease in the size of glaciers and ice sheets, a slight rise in the average ocean level, and more severe rains storms and tornados. In response to these phenomena, the Intergovernmental Panel on Climate Change (IPCC) gathered hundreds of scientists from around the world to study these problems. In a 2006 report, the IPCC stated that most of the observed increase in the global average temperature since the mid-20th century is very likely (greater than 90 percent) due to the observed increase in **anthropogenic greenhouse gas** concentrations. The three major gases are carbon dioxide (from pre-industrial levels of 280 ppm to 2003 levels of 380 ppm), **methane** (from pre-industrial levels of 715 ppb to 1774 ppb in 2005), and nitrous oxide (from pre-industrial levels of 270 ppb to 319 ppb in 2005). These gasses absorb the infrared heat radiating from the earth and thus heat the lower atmosphere. This warming is in addition to the normal warming of the atmosphere by the greenhouse effect. Review the diagram on page 194.

CO₂ and Temperature Graphs

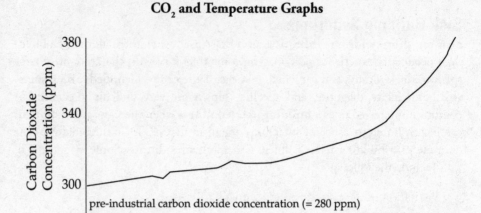

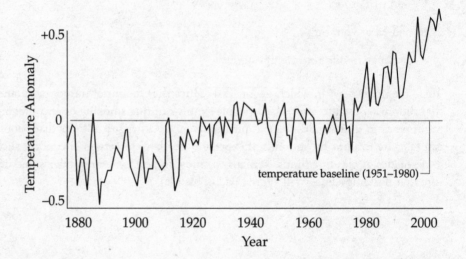

The increase in the earth's temperature will lead to a variety of changes to the earth. Physical changes on Earth include: further lessening of glaciers and ice sheets, continued rising of average ocean levels (due mostly to the thermal expansion of water), changes in precipitation patterns (with wet areas getting more precipitation and dry areas getting less precipitation), an increase in the frequency and duration of storms, an increase in the number of hot days, and a decrease in the number of cold days.

Climate change will also affect biota. While there will be increased crop yields in cold environments, this is likely to be offset by loss of croplands as other areas suffer droughts and higher temperatures. Cold-tolerant species will need to migrate to cooler climates or they may become extinct. Heat-tolerant species (including mosquitoes and other disease vectors) may spread and invade new habitats. Human health will show additional deaths from water and insect-borne diseases. More frequent heat spells will endanger the very young and old. It is very likely that commerce, transport facilities, and coastal settlements will be disrupted by ocean level changes and stronger and more frequent storms. Marine ecosystem productivity and fishery productivity is also likely to change.

Adaptations to the warmer climate will need to occur at many levels of society. Technological improvements like carbon sequestration and the reduction of emissions from engines, behavioral changes such as turning off lights to conserve electricity, and policy changes such as enacting new treaties (like the **Kyoto Accord**) and legislation will all be necessary in the next few decades. The promotion of sustainable growth will enhance the abilities of all societies to adapt to the new climate.

THERMAL POLLUTION

Urban environments are generally about 20 degrees (Farenheit) warmer than the countryside that surrounds them, and this is due to the heat absorbing capacity of buildings, concrete, and asphalt, which radiate the heat that they have absorbed. Industrial and domestic machines also directly warm the air. Because of their high temperatures, urban areas are known as **heat islands.** The high temperatures of heat islands increase the rates of photochemical reactions, which in turn leads to photochemical smog.

The temperature profile of an urban area shows peaks and valleys in temperature based on how the land is used. For example, green spaces have lower temperatures than commercial areas, which have lots of parking lots, cars, buildings, and asphalt. Two ways in which the heat island effect can be significantly reduced are: replace dark, heat-absorbing surfaces (such as roofs) with light-colored heat-reflecting surfaces; and plant trees and add to green spaces. Trees shade the urban environment from solar radiation; in addition, the process of transpiration (the release of water through plant leaves) creates a cooling effect for the surround area. Another reason why urban areas are often less cool than rural areas is that the concrete and asphalt in cities increase water runoff. Runoff leads to increased temperatures because the deep pools of water that are created as a result of runoff are less affected by evaporation than are areas where water is spread out thinly over a larger surface area. Green spaces can reduce runoff by trapping the water and distributing it more evenly across a larger surface area.

One way people are trying to combat thermal pollution is by adding green roofs to city buildings. A green roof, or living roof, is a roof that is fully or partially covered with plants, greenery, gardens, and other vegetation planted over some type of waterproofing material. Not only does this combat the heat island effect, but it also keeps the buildings cool in summer and warm in winter, reduces rainwater runoff, provides habitats for wildlife, and helps to clean the urban air.

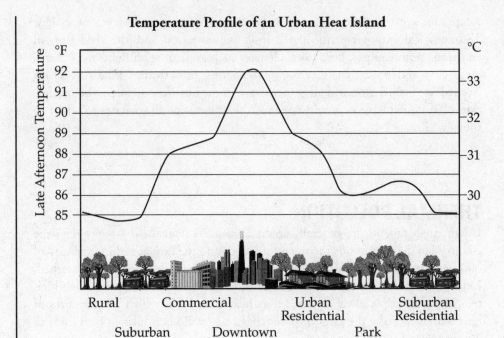

Temperature Profile of an Urban Heat Island

Another type of thermal pollution associated with many urban environments is **temperature inversion**. In this phenomenon, air pollutants become trapped over cities because they are not able to rise into the atmosphere. In normal atmospheric conditions, the warm polluted air over a city rises into the cooler atmosphere. (Remember that warm air is less dense than the surrounding cool air, and less dense objects float!) In an inversion, the air above the city is warm, and blocks the polluted air from rising. The polluted air remains hanging above the city, and can cause respiratory problems. Inversions often occur in cities surrounded by mountain, or cities bordered by mountains on one side and ocean on the other (for example, Los Angeles and Beirut). But thermal inversions can occur over any city where large masses of warm air can become stalled.

Thermal Inversion

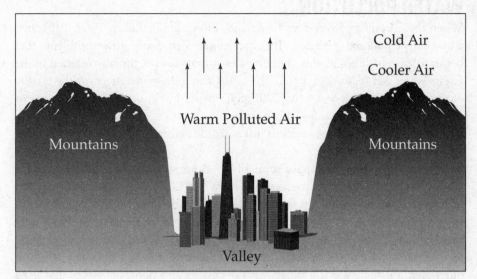

Normal Conditions

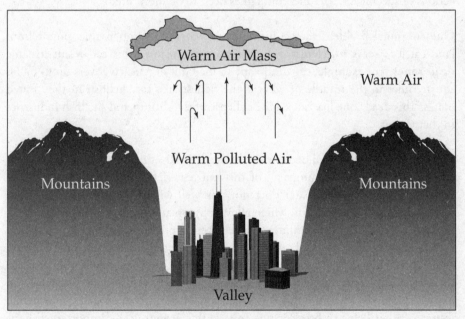

A Temperature Inversion

WATER POLLUTION

When the Cuyahoga River near Cleveland, Ohio caught fire in 1969, it became a symbol of polluted America. This fire, along with many other problems that began to arise with polluted bodies of water at that time, eventually resulted in the Clean Water Act (CWA) of 1972. The CWA had a dramatic effect on the quality of water in the United States. By 2006, 89.3 percent of community water systems met federal health standards—this number was up from the 79 percent that were considered clean by the government just a decade before.

Experts say that Americans have some of the cleanest drinking (tap) water in the world. From the time of the passage of the CWA to 2002, 60 percent of the stream lengths that were tested were found to be sufficiently clean to allow fishing and swimming, while only 36 percent of the streams that were tested in 1972 were clean enough. Also as a result of the CWA, the annual loss of wetlands decreased by 80 percent from 1972 to 2002. The CWA has certainly had a positive effect on our water, but there are still plenty of water issues and bodies of water that need to be cleaned; plus, the Clean Water Act needs to be constantly enforced, and the actions of specific citizens and companies need to be monitored.

One continual problem that contributes to water pollution is that runoff from land carries **excess nutrients** and pollutants to streams. This can result in large **dead zones**. For example, the dead zone in the Gulf of Mexico covers up to 5,000 square miles in the middle of what is the richest area for shellfish in the United States. This dead zone has caused the collapse of the shrimp and shellfish industries in that region.

The dead zone was created because the Mississippi River collects runoff as it travels through farmlands and dumps all of this nutrient-rich water into the Gulf. The warm, nutrient-rich freshwater does not mix well with the colder saltwater, and this results in **eutrophication**, which allows phytoplankton to grow almost uncontrollably. In turn, the zooplankton that feed on them also experience a population explosion. When the phytoplankton and zooplankton die and sink to the bottom, bacteria metabolize the available dissolved oxygen as they decompose this detritus; the lack of oxygen creates a **hypoxic zone**, in which nothing that depends on oxygen can grow. This zone stays in place from May until September when colder, wetter weather helps to break it up. To save this economically important fishery, Congress has introduced a plan of action that should reduce the size of the dead zone by 50 percent by the year 2015.

You should review the diagram of the Mississippi dead zone given on the following page. The black areas (near the coast of Louisiana, Alabama, and Texas) represent areas where the level of dissolved oxygen (DO) is very low.

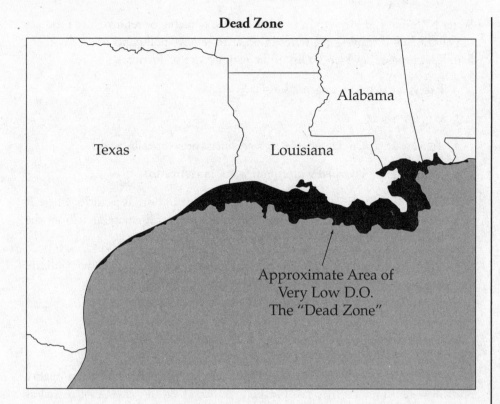

Dead Zone

Alabama

Texas

Louisiana

Approximate Area of
Very Low D.O.
The "Dead Zone"

Sources

Like the terms that are used to describe sources of air pollution, particular sources that are responsible for water pollution, like paper mills, are called **point sources,** and pollution that does not have a definitive source (or results from contributions of many sources) are **nonpoint sources.**

Right now, the biggest source of water pollution is agricultural activities; the runners-up are industrial and mining activities. Unfortunately, standing bodies of water such as ponds, reservoirs, and lakes do not recover quickly from the addition of pollutants. The lack of water flow prevents the pollutants from being diluted, which means that they accumulate in the water and undergo biomagnification in the food chain. In a similar way, groundwater does not recover well from the addition of pollutants; this is again because there is very little movement of water and therefore very little flushing, mixing, or dilution. Furthermore, groundwater is generally very cold and low in dissolved oxygen, which makes recovery from degradable waste a slow process. The porous rock that surrounds the groundwater absorbs the pollutants, which makes them difficult to remove.

However, flowing streams and rivers can recover from moderate levels of pollutants if the pollutants are degradable. As illustrated by the implementation of long sewage pipes that once dumped raw sewage into the ocean off coastal areas, people thought that the ocean was able to dilute and recover from the addition of any amount of pollutants. While oceans can dilute, flush, and decompose large amounts of degradable waste, their capacity for recovery is unknown.

Water pollution is dealt with in two basic ways: reducing or removing the sources of pollution, and treating the water in order to remove pollutants or render them harmless in some way. Here's a list of the major water pollutants.

- Excess nutrients (nitrogen, phosphate, etc.)

- Organic waste

- Toxic waste (pesticides, petroleum products, heavy metals, acids)

- Sediments (soil washed with runoff water into streams)

- Hot or Cold water (hot water discharged from industrial facilities where it was used as a coolant; cold water from dam releases discharging it from the bottom of a reservoir)

- Coliform bacteria (bacteria found in the intestines of animals that indicate the presence of fecal matter in water)

- Invasive species (zebra mussels, for example)

- Thermal pollution

Finally, perhaps your AP Environmental Science class has performed water quality tests on water samples. They test for the presence of various chemicals as well as insect larvae, which act as indicator species. Among the most important factors in judging water quality are

- **pH,** which is a measure of acidity or alkalinity (normal for water is 6–8)

- **Hardness,** which is a measure of the concentrations of calcium and magnesium

- **Dissolved oxygen**—low levels of dissolved oxygen indicate an inability to sustain life (warm water holds less dissolved oxygen than cool water)

- **Turbidity**—or the density of suspended particles in the water

- **BOD,** which is a measure of the rate at which bacteria absorb O from the water.

Now, let's talk more specifically about a major water pollutant—wastewater.

Wastewater

Another group of water pollutants that are very dangerous to human health are infectious agents, such as those found in human and animal waste. Fecal waste not only contains the symbiotic bacteria that aid in the human digestive processes, it also contains disease-causing bacteria. Several human diseases, such as cholera and typhoid fever, are caused as a result of human waste entering the water source of a community. In fact, the major reason for the increase in the life span of humans was not modern developments in medicine; it was the introduction of cleaner drinking water and better ways of disposing of wastewater.

The term **wastewater** is used to refer to any water that has been used by humans. This includes human sewage; water drained from showers, tubs, sinks, dishwashers, and washing machines; water from industrial processes; and storm water runoff. Water that is channeled into storm drains, such as storm water, is generally dumped directly into rivers. (This is why storm drain covers in many locations have been stenciled with warnings about not dumping material down them.)

Today in the United States, wastewater that isn't storm water is moved through sewage pipes to a sewage treatment facility, but this was not always the case. Sewage water once was, and in developing countries still is, merely dumped into the nearest river or ocean. While some amounts of sewage can be diluted and broken down in these waters, too much waste poses serious risks to human health and the health of the aquatic ecosystems.

Now in the United States, sewage pipes deliver wastewater to a municipal sewage treatment plant, where it is first filtered through screens (in what's called a **physical treatment**) to remove debris such as stones, sticks, rags, toys, and other objects that were flushed down the toilet. This debris is then usually separated and sent to a landfill. The remaining water is passed into a settling tank, where suspended solids settle out as sludge—chemically treated polymers may be added to help the suspended solids separate and settle out. This treatment is known as **primary treatment** and it removes about 60 percent of the suspended solids and 30 percent of the organic waste that requires oxygen in order to decompose.

Secondary treatment refers to the biological treatment of the wastewater in order to continue to remove biodegradable waste. This treatment can be done using trickling filters, in which aerobic bacteria digest waste as it seeps over bacteria-covered rock beds. Alternately, the wastewater can be pumped into an activated **sludge processor,** which is basically a tank filled with aerobic bacteria. The solids in the water, including the bacteria, are once again left to settle out. The solids remaining are considered **sludge,** which is combined with the sludge from the primary treatment. Sludge used to be dumped into the ocean, but that practice has been banned. Instead, the sludge is further processed with anaerobic bacteria (to breakdown more organic material). This digestion also produces methane gas that can be used as an alternative fuel to run the treatment plant. After drying, this sludge cake can be processed and sold as fertilizer.

At the end of secondary treatment, 97 percent of the suspended solids; 95–97 percent of the organic waste; 70 percent of the toxic metals, organic chemicals, and phosphates; 50 percent of the nitrogen; and 5 percent of the dissolved salts have been removed from the wastewater. However, almost no persistent organic chemicals, such as pesticides, are removed, nor are radioactive isotopes. Generally, after secondary treatment, the wastewater is chlorinated to remove any remaining living cells and then discharged into a stream, the ocean, or water that's used to water lawns (called **gray water**). A negative effect of the final chlorination of the water is that trihalomethanes (potential carcinogens) can be formed when any organic matter left in the water reacts with the chlorine, and this is problematic. Two alternate processes to chlorination—ozonation and UV radiation—have been used to treat secondary-treatment water, but they have not proven to be as effective or long-lasting as chlorine, and are also much more expensive.

Some municipal plants deposit wastewater directly into ground water; this is done in San Jose Creek in Los Angeles County. In these places, the water must be further treated by tertiary treatment. **Tertiary treatment** involves passing the secondary treated water through a series of sand and carbon filters, and then further chlorination. At the San Jose Creek Plant, the tertiary treated water from the reclamation plants is discharged into percolation basins, where it replenishes groundwater, or it is used for irrigation and for watering lawns, golf courses, and plants in nurseries. Tertiary treatment is expensive, but in arid or semi-arid regions, every gallon that can be reclaimed is one that need not come from rapidly depleting sources, such as diminished rivers or underground aquifers.

Private wastewater treatment in the form of septic tank systems is hallmarked by some as the most environmentally friendly type of waste disposal. Septic tanks act in a way that's similar to the primary and secondary treatments that take place in municipal treatment plants. The water is then discharged into leachate (drain) fields. In order to install these types of systems, the soil must be able to percolate the water—that is, the water must be made to move from the top of the soil though its various horizons. Some clay soils are not porous enough to allow percolation and thus are unsuitable for a septic field.

Water Quality Legislation

There are many pieces of federal law that cover water quality. Be sure that you are familiar with these laws for the exam.

Date	Name of Legislation	What It Did
1972	Clean Water Act	Used regulatory and non-regulatory tools to protect all surface waters in the United States. • Sharply reduced direct pollutant discharges into waterways • Financed municipal wastewater treatment facilities, and manages polluted runoff • Achieved the broader goal of restoring and maintaining the chemical, physical, and biological integrity of the nation's waters • Supported "the protection and propagation of fish, shellfish, and wildlife, and recreation in and on the water"
1972	Ocean Dumping Act	Made it unlawful for any person to dump, or transport for the purpose of dumping, sewage, sludge, or industrial waste into ocean waters.
1974, 1996	Safe Drinking Water Act	Established a federal program to monitor and increase the safety of the drinking water supply. It does not apply to wells that supply fewer than 25 people.
1990	Oil Spill Prevention and Liability Act	Strengthened EPA's ability to prevent and respond to catastrophic oil spills. Established a trust fund (financed by a tax on oil), which is available to clean up spills.

SOLID WASTE (GARBAGE)

Solid waste can consist of hazardous waste, industrial solid waste, or municipal waste. Many types of solid waste provide a threat to human health and the environment.

The phrase "reduce, reuse, recycle" might seem simplistic, but it does outline the steps needed to reduce the amount of solid waste that must be dealt with. "Reduce," of course, refers to the minimizing of disposable waste. There are many types of packaging that are extremely wasteful—if you keep an eye out, you'll see them everywhere. "Reuse" applies to products that in some cases are disposable, but in other forms can be used over and over again, such as refillable bottles and tanks, reusable packing materials, secondhand goods, and cloth shopping bags. Reusing products prevents these high-quality goods from becoming waste. Finally, "recycling" is the reuse of materials. In **primary recycling,** materials such as plastic or aluminum are used to rebuild the same product—an example of this is the use of the aluminum from aluminum cans to produce more aluminum cans. Alternately, in **secondary recycling,** materials are reused to form new products that are usually lower quality goods—examples of this are old tires are recycled to form carpet, and plastic bottles are recycled to create decking material. Finally, another environmentally important process is composting. **Composting** allows the organic material in solid waste to be decomposed and reintroduced into the soil.

According to the EPA, one of the most effective steps in aiding the environment that occurred in the twentieth century was the marked growth in the use of recycling and composting to deal with solid waste. Although there are still some products that are not feasibly recycled, those that were either recyclable or suitable for composting diverted more than 86.6 millions tons of material away from landfills and incinerators in 2012! That's nearly three times the amount that was recycled or composted in 1990. According to the EPA, the following percent of each of these materials was recycled in the year 2012:

Material	Percent of all of this material that was recycled in 2003
Lead-acid batteries	95.9
Steel cans	70.8
Newspaper	70.0
Yard trimmings	57.7
Soda cans	54.6
Tires	44.6
Glass	34.1
Pet bottles	30.8
Electronics	29.2

In order to encourage people to reduce, reduce, and recycle, many communities have established Pay-As-You-Throw (PAYT) programs, which charge municipal customers for the amount of household garbage they throw away. As you can imagine, this has been a strong incentive for people to practice these good habits. An example of an incentive program that has worked beautifully is the bottle redemption bill. Ten states have enacted bottle bills and they have worked fantastically, especially in Michigan, which has a ten cent redemption fee.

Landfills

In 1987, after it was discovered that landfills on Long Island were contaminating local groundwater, the barge Mabro left New York towing 3,186 tons of garbage, in search of a dumping ground. However, it was barred from docking in several southern states, and then the countries of Mexico, Cuba, and Belize. Three months and 6,000 miles later, it returned to New York, where it became a symbol for Americans who were concerned about the status of landfills in the

United States. It was also at this time that the term NIMBY (which stands for Not In My Backyard) became popular. It was widely agreed upon that landfills were needed, but no one wanted a landfill close to their home.

Modern landfills are very different from the traditional caricature of a garbage dump filled with heaps of junked cars and rats foraging for food scraps. Federal regulations that protect human health and the environment have paved the way for **sanitary landfills**. For example, federal law prohibits landfills from being located near geological faults, wetlands, or flood plains. Additionally, landfill sites are periodically required to dig large holes in the ground and line them with geomembranes or plastic sheets that are reinforced with two feet of clay on the bottom and sides. Smoothing wet clay is much like making a clay pot; the layer that is created is virtually impermeable. Also, the waste in the landfill must be frequently covered with soil in order to control insects, bacteria, rodents, and odor; and the decomposed material that percolates to the bottom of the landfill (called **leachate**) is piped to the top of the site and collected in leachate ponds, which are closely monitored. Gases from the landfill, like methane, may even be piped up from the site and used to generate electricity. Sometimes the methane is burned in continuously flaming flares to avoid larger fires or explosions. To ensure that landfills do not contaminate the environment, they are required to be positioned at least six feet above the water table, and groundwater at the sites must be tested frequently for quality. When one site (hole) is full, it must be capped with an engineered cover, monitored, and provided with long-term care.

Waste may also be burned in municipal incinerators, which are generally capable of sorting out recyclables first. The energy released from the incineration can be used to generate electricity in what's called the **Waste-to-Energy (WTE) program.** This type of system is particularly effective in large municipal areas, where waste only needs to be transported short distances.

HAZARDOUS WASTE

Hazardous waste is any waste that poses a danger to human health; it must be dealt with in a different way than other types of waste. Hazardous waste includes such common items as batteries, cleaners, paints, solvents, and pesticides. Industry produces the largest amounts of hazardous waste, and most developed countries now regulate the disposal of these wastes. United States law mandates that hazardous materials be tracked "from cradle to grave" thanks to laws like The Resource Conservation and Recovery Act (RCRA). The EPA breaks hazardous wastes down into four categories.

- **Corrosive waste:** Waste that corrodes metal

- **Ignitable waste:** Substances such as alcohol or gasoline that can easily catch fire

- **Reactive waste:** Substances that are chemically unstable or react readily with other compounds, resulting in explosions or causing other problems

- **Toxic waste:** Waste that creates health risks when inhaled or ingested, or when it comes into contact with skin

Hazardous wastes are disposed of in three main ways: in injection wells, in surface impoundments, and in landfills. Many communities have specific areas in their landfills that are designated for hazardous waste, and the standards for those areas of the landfills are higher than standards for non-hazardous waste areas. **Surface impoundment** is typically used for liquid waste; it involves the creation of shallow, lined pools from which the hazardous liquid evaporates. **Deep well injection** involves drilling a hole in the ground that's below the water table. These wells must reach below the impervious soil layer into porous rock, and waste is injected into the well. All three of these methods have their advantages, but none of them is satisfactory.

As you can probably imagine, radioactive waste must be contained in a different way than other hazardous wastes. For years, the United States has been trying to develop one major site for the disposal of all of our radioactive waste. Yucca Mountain, Nevada was selected as the location of this site because of its remoteness and because nuclear testing had previously been done at the site. This decision is still controversial, in part, because of NIMBY, but also because Nevada has no nuclear power plants, so why should they be responsible for the spent fuel repository? It's a contentious subject. However, it is critical that some plan for the long-term storage of nuclear waste is made, because in the very near future the nation's nuclear power plants will be at maximum capacity for the storage of spent fuel. The Waste Isolation Pilot Plant in New Mexico is a new permanent site for **nuclear waste burial**. Waste that's left over from the construction of nuclear weapons is known as **transuranic waste**.

Some people define radioactive wastes that produce low levels of ionizing radiation as **low-level radioactive waste** and those that produce high levels of ionizing radiation as **high-level radioactive waste.** However, the EPA categorizes radioactive waste according to its place of origin. Therefore, in the EPA's classification system, some wastes that are considered high level may actually be less radioactive than certain low-level wastes. The EPA puts all radioactive wastes into six categories.

- Nuclear reactor waste: high level

- Waste from the reprocessing of spent nuclear fuel: high level

- Waste from the manufacture of nuclear weapons: high level

- Waste from the mining and processing of uranium ore: high level

- Radioactive waste from industrial or research industries, including clothing, gloves, tubes, needles, animal carcasses, etc.: low level

- Radioactive natural materials: not a waste

In general, low-level waste is either stored on-site by licensed facilities until the radioactivity has degraded, or it is shipped to a low-level waste disposal facility. Mixed waste, containing both chemically hazardous waste and radioactive waste, is generally disposed of in the same manner.

In this book, we use the radioactive waste disposal terms used in the EPA classification system. However, on the AP Environmental Science Exam, a discussion of either system of classification would be considered correct—as long as you identify which classification system you're using.

Contaminated Waste Sites

After the 1970s, new regulations for the disposal of hazardous wastes solved many of the problems of how to add new wastes to landfills with minimal impact on the environment; at the same time, the issue lingered of what to do with sites that were already problematic. These sites had to be cleaned up and those who had acted irresponsibly had to be held accountable for the environmental problems they'd caused. For these reasons, the United States legislature created the **Superfund Program**, which was administered by the EPA.

Rocky Flats, Colorado is a Superfund site where the party responsible for the damage happened to be the United States government. Starting in 1952 and continuing for almost 40 years, components of nuclear weapons, such as plutonium, uranium, beryllium, and stainless steel were all manufactured on this site. Now that the area has been significantly cleaned up, it is home to a variety of plants and animals, including bald eagles, and acts as a wind-power testing site.

Along with the Rocky Flats, you should know the story of **Love Canal** near Niagara Falls, NY. The site was originally a canal built to bring power and employment to the surrounding community. After the canal's failure, the land was purchased by various companies that turned the canal into a landfill. After the town purchased the covered landfill area, 100 homes and a school were built on the site. In 1978 people saw rusting drums full of waste sticking up above ground. They also noticed dead and dying trees and gardens. Homeowners even reported having pools of smelly liquids in their basements; their children reported burning hands and faces after coming in from playing. Environmental Protection Agency employees soon came to the canal area, and by the end of August, 220 families had moved or said they would move out of the area. It was in response to the situation at Love Canal (and other cites in the United States) that laws like The Resource Conservation and Recovery Act (RCRA) and The Comprehensive Environmental Response, Compensation, and Liability Act (CERCLA) were passed.

Laws for Solid and Hazardous Wastes

There are many federal statutes that cover issues concerning solid and hazardous wastes. You should be familiar with the ones given below.

Date	Law	What It Did
1975	Hazardous Materials Transportation Act	Governed the transportation of hazardous material and wastes in commerce.
1976	The Resource Conservation and Recovery Act	• The solid waste program encouraged states to develop comprehensive plans to manage nonhazardous industrial solid waste and municipal solid waste; sets criteria for municipal solid waste landfills and other solid waste disposal facilities; and prohibits the open dumping of solid waste. • The hazardous waste program established a system for controlling hazardous waste from the time it is generated until its ultimate disposal—in effect, from "cradle to grave." • The underground storage tank (UST) program regulates underground storage tanks containing hazardous substances and petroleum products.
1976	Toxic Substances Control Act (TOSCA)	Gave the EPA the ability to track the 75,000 industrial chemicals currently produced in or imported to the United States. EPA repeatedly screens these chemicals and can require reporting or testing of those that may pose an environmental or human-health hazard. Allows the EPA to ban the manufacture and import of those chemicals that pose an unreasonable risk.
1980	The Comprehensive Environmental Response, Compensation, and Liability Act (CERCLA), commonly known as Superfund	• Created a tax on the chemical and petroleum industries and provided broad Federal authority to respond directly to releases or threatened releases of hazardous substances that may endanger public health or the environment. • Established prohibitions and requirements concerning closed and abandoned hazardous waste sites. • Provided for liability of persons responsible for releases of hazardous waste at these sites. • Established a trust fund to provide for cleanup when no responsible party could be identified.
1982	Nuclear Waste Policy Act	Established both the Federal government's responsibility to provide a place for the permanent disposal of high-level radioactive waste and spent nuclear fuel, and the generators' responsibility to bear the costs of permanent disposal.

NOISE POLLUTION

Take your earphones off and think about this for a minute: The EPA considers noise to be a controllable pollutant. The **U.S. Noise Control Act** of 1972 gave the EPA power to set emission standards for major sources of noise, including transportation, machinery, and construction. Occupational Safety and Health Association (OSHA) has also set limits on the amount of noise that people can be exposed to in the workplace. Although the definition of noise pollution can be quite flexible, **noise pollution** in a broad sense is any noise that causes stress or has the potential to damage human health.

One concern about noise is that continued exposure to high levels can damage hearing. The louder the noise, the shorter the exposure it takes to damage inner ear cells and cause hearing impairment. Unfortunately, certain essential cells in the ear that are involved in hearing do not regenerate: so the loss of hearing is permanent. There are federal laws that regulate noise emissions from some equipment and modes of transportation, and OSHA is responsible for the regulation of noise in the workplace; in local communities, however, noise pollution is usually controlled by state or local laws. The **Quiet Communities Act** provides for the coordination of federal research and activities into noise control. Furthermore, this act authorized the use of Federal Aviation Administration funds for the development of noise abatement plans around airports.

All right, we're finished explaining pollution. You have two more chapters to get through and then you'll be ready to take the practice exam. In the next chapter, we discuss aspects of culture and society that are pertinent to the exam. First, study the key terms. Then, complete the drill and read carefully through the answers and explanations—and move on: you're almost done!

KEY TERMS

Don't *waste* any time; study these words now!

Toxicity & Health

toxicity
toxin
dose-response analysis
dose-response curve
LD_{50}
ED_{50}
poison
threshold dose
acute effect
chronic effect
infection
disease
pathogen
vector
risk
risk assessment
risk management

Air Pollution

primary and secondary pollutants
sulfur oxides
nitrogen oxides
CO and CO_2
stationary sources
moving sources
point source pollution
non-point source pollution
criteria pollutants
tropospheric ozone vs stratospheric ozone
industrial smog (grey smog)
photochemical smog (brown smog)
chlorofluorocarbons (CFCs)
ozone loss
acid precipitation
dry acid particle deposition
wet deposition

National Ambient Air Quality Standards

catalytic converter
CAFE
VOCs
sick building syndrome

Climate Change

methane
anthropogenic greenhouse gases
carbon sequestration
Kyoto accord
thermal pollution
heat island
temperature inversion

Water Pollution

excess nutrients
dead zone
eutrophication
hypoxic zone
pH
hardness
dissolved oxygen
turbidity
BOD
wastewater
physical treatment
primary, secondary, tertiary treatment
sludge
sludge processor
gray water

Solid Waste

secondary recycling
primary recycling
composting
sanitary landfill
leachate
Waste-to-Energy programs

Hazardous Waste

surface impoundment
corrosive, ignitible, reactive and toxic wastes
deep well injection
nuclear waste burial
transuranic waste
low-level, high-level radioactive waste
Superfund Program
Love Canal

Noise Pollution

U.S. Noise Control Act
noise pollution
Quiet Commonalities Act

Chapter 9 Drill

Directions: Each of the questions or incomplete statements below is followed by five suggested answers or completions. Select the one that is best in each case. For answers and explanations, see Chapter 13.

1. Which of the following is NOT a direct source of groundwater pollution?

   (A) Automobile exhaust
   (B) Wastewater lagoons
   (C) Underground storage tanks
   (D) Waste injected into deep wells
   (E) Pesticides sprayed on the land

2. Which of the following cities would have the greatest amount of gray-air smog?

   (A) New York, New York
   (B) Beijing, China
   (C) Los Angles, California
   (D) Chicago, Illinois
   (E) London, England

3. All of the following are true about sanitary landfills EXCEPT that they

   (A) have methods of monitoring leaks in the clay and plastic liners
   (B) pipe generated methane gas to storage tanks
   (C) pump leachate out of the landfill for treatment and disposal
   (D) are built so that trash sits on top of the land
   (E) contain clay and plastic liners that prevent leachate from entering the soil

4. Which of the following correctly explains what happens to the level of oxygen dissolved in water when organic waste is put in the water?

   (A) The levels would remain the same after the waste was
   added.
   (B) The levels would increase because of the availability
   of nutrients to animals that live in the water.
   (C) The levels would increase because of the higher temperatures of the water.
   (D) The levels would decrease because of the waste absorbing the oxygen.
   (E) The levels would decrease because of the bacteria feeding off the waste and using the oxygen to live.

5. An abundance of which of the following creatures (and a lack of other creatures) would indicate that water is polluted?

   (A) Trout and other game fish
   (B) Sludge worms, anaerobic bacteria, and fungi
   (C) Carp, gar, and leeches
   (D) Salamanders and turtles
   (E) Insect larvae and nymphs

6. Which of the following is the most common way of disposing of municipal solid waste?

   (A) Recycling
   (B) Composting
   (C) Placing in landfills
   (D) Burning
   (E) Transporting to other countries

7. Which of the following gases involved in global climate change is increasing in the atmosphere at the fastest rate?

   (A) H_2O
   (B) Methane
   (C) Chlorofluorocarbons
   (D) CO_2
   (E) O_2

8. Which of the following choices gives the correct order of processing sanitary waste in a sewage treatment plant?

   (A) Disinfection—breakdown of organics by bacteria—solid separation
   (B) Solid separation—breakdown of organics by bacteria—disinfection
   (C) Solid separation—disinfection—breakdown of organics by bacteria
   (D) Breakdown of organics by bacteria—solid separation—disinfection
   (E) Breakdown of organics by bacteria—disinfection—solid separation

9. Which of the following is a secondary pollutant?

   (A) CO
   (B) Soot
   (C) VOCs
   (D) PANs
   (E) CO_2

10. The United States was building a nuclear waste disposal site in

   (A) Wheeling, West Virginia
   (B) Yucca Mountain, Nevada
   (C) Gallup, New Mexico
   (D) Hudson, New York
   (E) Johnstown, Pennsylvania

11. Oxides of nitrogen create the pollutant

   (A) nitric acid
   (B) nitrogen gas
   (C) sulfuric acid
   (D) carbonic acid
   (E) phosphoric acid

12. In comparison to the surrounding rural areas, cities are

   (A) cooler than the rural area
   (B) the same temperature as the rural area
   (C) hotter than the rural area
   (D) incomparable to the surrounding areas as far as temperatures
   (E) more likely to have snowfall than the surrounding area

Directions: Each set of lettered choices below refers to the numbered questions or statements immediately following it. Select the one lettered choice that best answers each questions or best fits each statement. A choice may be used once, more than once, or not at all in each set.

Questions 13-17 refer to the following methods of treating hazardous wastes.
   (A) underground burial
   (B) reduce the amount made
   (C) incineration
   (D) neutralization
   (E) bioremediation

13. Involves mixing waste with other chemicals to produce less toxic substances

14. This process can cause secondary pollutants in the atmosphere

15. Injection into deep wells

16. The easiest and cheapest method

17. Living organisms process the waste and remove it from the ecosystem

Free-Response Question

1. Photochemical smog is one of the most common forms of air pollution today.

    (a) Identify two primary pollutants that cause photochemical smog. Describe how they are produced.
    (b) Identify two secondary pollutants that make up photochemical smog. Describe how they are produced.
    (c) Give one reason why photochemical smog is more likely to be found in industrialized nations and gray-air smog is more likely to be found in non-industrialized nations.
    (d) Give one component of photochemical smog that affects human health. Explain its consequences.

REFLECT ACTIVITY

Respond to the following questions:

- For which content topics discussed in this chapter do you feel you have achieved sufficient mastery to answer multiple-choice questions correctly?

- For which content topics discussed in this chapter do you feel you have achieved sufficient mastery to identify hot-button topics and successfully articulate them in free-response questions?

- For which content topics discussed in this chapter do you feel you need more work before you can answer multiple-choice questions correctly?

- For which content topics discussed in this chapter do you feel you need more work before you can successfully identify and articulate them in free-response questions?

- What parts of this chapter are you going to re-review?

- Will you seek further help, outside of this book (such as a teacher, tutor, or AP Central), on any of the content in this chapter—and, if so, on what content?

Chapter 10
Culture, Society, and Environmental Quality

> Conservation is a great moral issue, for it involves the patriotic duty of ensuring the safety and continuance of the nation.
>
> —Theodore Roosevelt

This chapter contains the odds and ends about cultures, societies, and other topics that will be tested on the AP Environmental Science Exam, for which there are no specific headings in the course outline.

By now, you've probably almost completed a course in AP Environmental Science, which means that you realize that what's good for the environment is usually good for people because we live in an ecosystem on Earth just as all other plants and animals do. At this point, the idea of pursuing environmentally thoughtful courses of action, through government policy as well as our day-to-day actions, must seem like a great idea, right? Well, if environmental protection is good, why do we often get the message that it will cost us more, that people will lose jobs, or that it will infringe on our liberties in some way? The answer is that, to a certain extent, all of these things *will* happen. However, the health of humankind is dependent on the health of the earth, and often certain sacrifices must be made for the greater good of the earth.

In this chapter, we will first review the importance of **sustainability**. We will then move on to discuss public policy making, give you a brief history of environmental activism in the United States, and then go through the important acts and amendments that you'll need to know for the test. We'll wrap up the chapter with a discussion of green taxes—and then we'll be done. Let's begin!

THE IMPORTANCE OF BEING SUSTAINABLE

What does the term "sustainable" mean to you? How much would you be willing to sacrifice in order to sustain environmental quality? These are questions that all citizens should ask themselves before entering the voting booth.

To environmentalists, sustaining environmental quality usually means working in the biotic and abiotic environments in a way that ensures they are capable of functioning sustainably. However, along with maintaining a sustainable environment, maintaining the health and happiness of the human species would also be a part of most environmentalists' goals for Earth. The human species cannot exist in a nonsustainable environment; after all, humans are part of a larger ecosystem, just as all other species of living things are. Our advantage, however—or rather, our responsibility—lies in the fact that we are the most technologically advanced and capable species on the planet. We are also the ones causing the most damage.

As environmentally literate, reasonable citizens, we know that we're sometimes obliged to make choices that may not make everyone happy, but we strive to make choices that will ultimately benefit the greatest possible number of people.

The United States comprises only about 5 percent of the world population but consumes about 40 percent of the world's total resources, including about 30 percent of the world's energy resources. If every country consumed global resources to this extent, we would need more resources to live than Earth can supply. This is because most of the resources that we rely upon are limited—recall the fossil fuels we burn, the way that we use water, and the rate at which we produce and dispose of waste.

Public Policy

The exploitation of public resources has been the motivation behind environmental policy at the international, national, state, and local level for as long as public policies have been made. Strictly speaking, **policy** is defined as a plan or course of action—as of a government, political party, or business—intended to influence and determine decisions, actions, and other matters.

While policies that we make as a nation are usually fairly easy to enforce—because they often have our collective best interests as a nation in mind—international policy, as is established through the United Nations (UN), is only achievable and realistic if the affected countries all cooperate with decisions that are made collectively. For example, in the 1994 International Conference on Population and Development (which was sponsored by the UN), one of the goals agreed on by the participants at the conference was to enroll 90 percent of all boys and girls in primary school by 2010. However, this policy can be put into action only if the countries that signed the agreement are willing to carry out the necessary steps.

There are ways in which the UN can attempt to force countries to follow mandates that are agreed on by the majority; these include withholding borrowing power through the World Bank; trade rules; and withholding aid. However, there are often certain environmentally significant countries that don't belong to the UN, didn't sign whatever agreement is at issue, or that just don't have the infrastructure to enforce the objective—however worthy. Additionally, international agencies often don't have the power to control what happens inside a particular country.

Much more effective are international policies that are put into effect through treaties that the countries involved have all agreed to; their governments have all ratified the treaty. Obviously, policies that countries agree to are most often ones that benefit these countries in some way, so it isn't too surprising that they are more readily enforced. In other words, international laws that are not agreeable are not usually followed because they don't provide countries with incentives. Moreover, it is not possible to punish countries that don't follow these policies.

As we touched upon above, it's understandably much easier to enforce laws and policies in the United States than it is for us to police the other nations of the world. In the United States, state and local laws have an effect on the environment, but if there is a conflict between state or local law and federal law, most times federal law will take precedence. However, in some cases states have legislated controls that are even stricter than those the federal law requires. In these cases, the state laws are the ones enforced, rather than the more lenient federal law. Additionally, some laws are passed and enforced regionally because of particular geographic needs; one example of this is the difference in water laws to the east and west of the Mississippi River.

East of the Mississippi River, water laws are based on the principle that the upstream consumers control the water but, by law, cannot impede or reduce its flow or change its quality. A number of lawsuits based on this premise have been filed and are currently pending. One example is the diversion of water from the Apalachicola-Chattahoochee-Flint (ACF) and Alabama-Coosa-Tallapoosa (ACT) river basins. The state of Georgia would like to divert water from these basins to supply the growing needs of the urban area of Atlanta. The states of Alabama, Tennessee, and Florida, which are downstream from the diversion, are concerned about the flow and quality of water that will reach them if this diversion project is carried out. This controversy is still in the court system.

While we're on the topic, Atlanta is a good example of a city whose ecological footprint is far larger than the resources that are available on the land it occupies. (Remember that an ecological footprint is the amount of resources available to support a population and absorb its wastes.) City lawmakers are rightfully concerned about water shortages in the near future.

On the other hand, water laws west of the Mississippi are based on **water rights**. West of the Mississippi, it's held that the person who first files a claim on a water resource has rights to the use of the water. The amount of water an individual with a water right has claim to each year is determined by water flow that year (and also by how much water the individual wants!). But, regardless of the amount of water present or the place where the right was claimed, the oldest person who made the water claim gets to use his share of water before anyone else can partake. Obviously, water rights in the West do not require sharing, as they do in the East.

Environmental Policy in the United States —A Short History

Although the first laws of the United States, such as those contained in the Constitution, do not mention the environment specifically, the Bill of Rights includes the Fifth Amendment, which prohibits the taking of private property for public use without just compensation. This has been interpreted to include the "preventing of serious public harm" by those who wish to take private property and do something on it that will affect the environment or those around them in a negative way. Basically, this means that your neighbor cannot decide to build a small nuclear power plant on his residential property because this would violate zoning laws.

Early laws did not mention the environment because when these laws were written, there was so much land and so many resources in the United States that it was unimaginable that they could ever be in danger of being used up. In fact, many early laws, such as the Homestead Act of 1862, encouraged the settlement and exploitation of western lands. Others, such as the Mineral Lands Act of 1866, encouraged the use of resources, and, unfortunately, this exploitative act is still in effect for many mining regulations. A few years later, the General Mining Act of 1872 was a federal law created to systematically oversee and control prospecting and mining for economic minerals, such as gold, platinum, and silver, on federal public lands.

Shortly after the Civil War, as people continued to migrate to the West, it was realized that the United States did not have an endless supply of land or resources. In fact, in order to preserve some of the lands in the West that were being very quickly settled, the first national park, Yellowstone National Park, was established in 1872. Further legislative action in 1891 created the forest reserves, which made these lands off limits to logging in order to protect the land from being overharvested and to maintain the existing watersheds. This legislation marked the beginning of the federal government assuming an environmentally protective role.

Political and Cultural Activism

During this time period there were several men who stood out as early environmental activists, including Henry David Thoreau (1817–1862). Thoreau's book, *Walden*, describes his retreat from society and the quiet years that he spent living on Walden Pond studying nature.

Another important writer and scientist of this time period was George Perkins Marsh (1812–1939), whose book *Man and Nature* helped the American public understand that there are limits to natural resources. His plan for the conservation of resources is the basis for many of the resource conservation principles that we try to adhere to today. Another early environmental advocate was John Muir, a nature preservationist who founded the Sierra Club in 1892. He led a campaign for the protection of lands from human exploitation and advocated low-impact recreational activities such as hiking and camping; these ideas did not become popular until the 1960s.

As far as political leaders, arguably the most environmentally active president in the history of the United States was Theodore Roosevelt (1858–1919). Roosevelt was interested from an early age in the workings of the environment and even began his own natural history museum as a child. Interestingly enough, that collection became a part of the founding collection for New York's American Museum of Natural History.

Roosevelt's term as president has been called the **Golden Age of Conservation** because of the many environmentally friendly laws and policies he put into effect. During his presidency (1901–1909), he increased area of national forest lands by 400 percent (up to 194,000,000 acres), establishing 150 new national forests and adding area to others. He established the first 51 bird reserves, signing the first one into existence by asking his advisors, "Is there any law which prohibits me from declaring this island a bird refuge?" When his advisors determined that there was not, he signed the bill with gusto, announcing, "Very well, then, I so declare it." Additionally, he established five national parks, including the Grand Canyon, four national games preserves, eighteen national monuments (established under the 1906 Antiquities Act), twenty-four reclamation projects, and seven conservation commissions.

Roosevelt also appointed the first chief of the United States Forest Service in the history of the United States, Gifford Pinchot (1865–1946). Pinchot applied the principles of sustainable harvest and multiple-use to wildlife protection, recreation, and resource extraction.

As you're probably well aware, the 1960s were a turbulent time in U.S. history, when the baby boomers born after World War II came of age and began expressing their opinions to the world. The book *Silent Spring*, written by Rachel Carson in 1962, awoke in many Americans an awareness of the state of the environment. The air was dirty, the water was polluted, and hazardous wastes were collecting in landfills all over the country. Also at this time, the Apollo space missions allowed Americans to see planet Earth from afar for the first time, and this popularized the term "spaceship Earth." Paul Ehrlich's 1968 book *The Population Bomb* warned of the myriad problems that would arise along with the quickly increasing human population, and an entertainer named John Deutschendorf took the stage name of John Denver and began to popularize the environmental movement through song.

A multitude of environmental laws and policies were initiated during the presidency of Richard Nixon. For example, the first Earth Day was celebrated on April 22, 1970. Also in 1970, Nixon signed into law the National Environmental Policy Act (NEPA); this act created the Council on Environmental Quality and required the submission of an environmental impact statement before any major federal action could be taken. One of Nixon's major environmental contributions was to consolidate two agencies that had environmental responsibilities into a bureau called the Environmental Protection Agency (EPA). Finally, two major legislative actions were enacted in this new era of environmental awareness in the United States. The first was the Clean Air Act of 1963, which we have mentioned many times in these pages, and the second was the Clean Water Act (introduced in 1972).

There is your brief history of environmental activity and activists in the history of the United States. On the following pages, we'll highlight some other important environmental laws that you should be aware of for this exam. The ones in bold are those that have had a particularly significant impact. Make flashcards of all of these, as well as the Clean Air Act of 1963 and the Clean Water Act, so that you know them cold for test day!

Some Major Environmental Policy Acts

As you just learned, a few pieces of legislation helped form the environmental policy of the United States. The table on the next page shows some of the legislation important in U.S. environmental policy. If you want to see specific laws that deal with particular problems like endangered species, clear water, or mine pollution, go back to those chapters!

Date	Name of Legislation	What It Did
1970	National Environmental Policy Act	Created the Council on Environmental Quality that resulted in the creation of the Environmental Protection Agency (EPA) from the consolidation of various environmental agencies. It also mandates that federal agencies prepare *environmental impact statements*.
1983	International Environmental Protection Act	Authorized the president to assist countries in protecting and maintaining wildlife habitats and in developing sound wildlife management and plant conservation programs. Special efforts should be made to establish and maintain wildlife sanctuaries, reserves, and parks; enact and enforce anti-poaching measures; and identify, study, and catalog animal and plant species, especially in tropical environments.
1990	Pollution Prevention Act	Designed to promote source reduction (stop pollution from being produced).
1990	Environmental Education Act	Established the Office of Environmental Education within the Environmental Protection Agency to develop and administer a federal environmental education program.

What Have We Done for Us Lately?

Although some environmental bills and amendments have been added since 1985, lately there has been a distinct anti-environmental movement influencing government actions. Large, established environmental groups, such as the Sierra Club, have declined in membership, which is an interesting litmus test of environmentalism in America. However, hope lies in the fact that new grassroots environmental organizations are currently growing throughout the United States. **Non-governmental organizations (NGOs)** like Greenpeace and the World Wildlife Fund also play a role in protecting the environment. Four environmental issues that are expected to take center-stage in the twenty-first century include

- climate change

- water shortages and water supplies

- population growth

- loss of biodiversity

Keep your eye out for discussions of these topics in the news, and listen critically to campaigning politicians to see where they stand on these issues. Such issues will prove to impact you personally in more and more ways as time goes by. The discussion of how to best reduce greenhouse gas emissions, for example, currently revolves around what's called **cap-and-trade policy**, an approach that provides economic incentives for limiting emissions of pollutants.

HOW ELSE DO WE MAKE ENVIRONMENTAL PROGRESS?

Most often, the United States government has approached environmental issues by passing "command and control" laws. These laws set limits on factors, such as the amounts of pollution that are allowable from various sources, and they establish penalties for those that go over the limits. Wildlife has always been protected by similar types of legislation. There is no doubt that these laws have led to cleaner air and water, as well as the conservation of soil and other natural resources. Endangered and threatened species have also both been protected, and some extremely endangered species have even been able to recover somewhat.

However, there have always been problems with the "command and control" approach. For example, consider the Endangered Species Act. Red-cockaded wood-peckers are endangered because the open forests with big, old pine trees have been replaced by forests with younger, smaller pines. Also, periodic natural fires, which historically kept the pinewoods open, have been suppressed because humans have settled in these areas. Periodic fires are needed to control the brushy understory and keep the pinewoods open. Creating yet another problem for the endangered bird, timber owners have been known to kill them in order to avoid preserving their habitat. However, it's very hard to prove what happens to these birds—are they being exterminated by landowners, or are they simply migrating elsewhere or declining in number for other reasons?

A green tax shift is a fiscal policy that lowers taxes on income, including wages and profit, and raises taxes on consumption, particularly the unsustainable consumption of non-renewable resources. Some taxes that could be lowered by the implementation of a green tax shift are payroll and income taxes, and the following is a list of taxes that could be implemented or, if currently in existence, increased:

- Carbon taxes on the use of fossil fuels

- Taxes on the extraction of mineral, energy, and forestry products

- License fees for fishing and hunting

- Taxes on technologies and products that are associated with substantial negative externalities

- Garbage disposal taxes

- Taxes on effluents, emissions, and other hazardous wastes

Green Taxes

There are other approaches that are more successfully used to continue environmental improvement without forcing the enactment of other types of command and control. Over time it has become clear that the act of punishing actions that hurt the environment is not nearly as effective as rewarding actions that help the environment. Since the 1970s, the United States has substantially increased taxes on labor and modestly increased taxes on income, while allowing actions that create pollution and cause resource depletion to remain largely untaxed. The result is that the tax system of the United States encourages resource depletion and discourages investments in machinery and labor. A worldwide discussion is taking place about how to move away from taxing "goods," such as investments and employment (activities we should be encouraging), and toward taxing "bads," like pollution, which we would like to discourage. Pollution taxes have now been embraced by a growing number of mainstream economists and policy makers, and are just one of a new group of taxes called **green taxes**.

Additionally, taxes on certain forms of consumption may occur through the feebate approach, in which additional fees are imposed on less sustainable products—such as sport-utility vehicles—and then pooled to fund rebates on more sustainable alternatives, such as hybrid electric vehicles.

The three main goals of green taxes are

- the generation of revenue to correct past pollution damage and reduce future pollution

- to change behavior

- to use the funds received from pollution taxes for restoration

In this scheme, taxes serve as policy tools, as well as a way to protect the environment.

Market permits are also being used somewhat successfully to encourage reduction in pollutants. Market permits are cap and trade permits and they work in this way: Companies are allowed to buy permits that allow them to discharge a certain amount of substances into certain environmental outlets. If they can reduce their discharge, they are allowed to sell the remaining portion of their permit to another company. Economically speaking, it is to a company's advantage to reduce its discharge and sell the remainder of its allowable discharge to another company. But perhaps a better idea is for the government to buy back the unused permits rather than have them sold to another industry; this would reduce the overall discharge.

Many people think that subsidies (which are giveaways or tax breaks on certain resources to encourage their use) are hurting the environment more than they're helping it; detractors think that subsidies only encourage the use of unsustainable products.

Note that all policies, treaties, and laws are important to our environment. However, for purposes of the AP Environmental Science Exam, you will probably only be asked questions concerning United States federal laws, which is why we suggested that you commit the table of acts you saw earlier in this chapter to memory. International treaties, summits, and policies such as those that are directed from the United Nations or one of its agencies will probably come up on future exams, but that isn't your problem.

Globalization

As you can imagine, our world is becoming more and more interconnected. Aircraft can fly around the world in about 24 hours; we have instant communication worldwide via phones, television, and the Internet. This is called **globalization,** and it affects society, the economy, and the environment. Positive effects can be seen in new economic opportunities, our expanded access to information, and the interactions of many societies. For example, grapes can be grown in Chile, shipped north, and be sold in your supermarket in less than a week. There are also several negative impacts of globalization. In certain parts of China, large piles of unusable electronic components have been creating water pollution problems as rainwater leaches out heavy metals. The rapid spread of emerging diseases, increased levels of air pollution and hazardous waste, and the loss of marine fish stocks are just a few more examples of globalization's negative impacts. Not to mention those grapes from Chile used an enormous amount of energy (and fossil fuel) and probably added to the pollution during their trip north!

Remember when you read about the Commons, resources owned by no one but accessible by everyone? This concept is important when we consider global access to those resources. Fresh water, clean air, ample supplies of fish, and access to fertile croplands are all examples of the global Commons. It is important to use these resources sustainably because they are the foundations for economic and social development.

Poverty and greed can cause people to use resources in an unsustainable manner and damage the environment. Cutting down important rainforest habitats to raise crops and accepting companies that generate a lot of harmful pollutants are two examples of how people's hunger for money can lead to unsustainable practices. Unfortunately, the economically disadvantaged people who allow unsustainable practices to continue are also the ones most susceptible to environmental issues brought about by climate change, and have the least amount of resources to combat the health and environmental problems that result.

International organizations such as the World Bank and the United Nations are two examples of institutions that are trying to ameliorate the poverty issue. The World Bank uses loans to reduce poverty and to help foster improvements in biodiversity, environmental policies, land management, pollution management, and water resources management. The United Nations, through its environmental program, seeks to promote international cooperation, develop regional programs to promote sustainability, and to assess global, regional, and national environmental trends.

There are several international agreements that cover pollution issues. Review some of them in the table below.

Date	Name of Agreement	What It Did
1978	Montreal Accord	Cut the emissions of CFCs that damage the ozone layer. This was amended in Copenhagen (1992) to include other key ozone-depleting chemicals.
1992	Basel Convention on the Control of the Transboundary Movements of Hazardous Wastes	169 parties aimed to protect human health and the environment against the adverse effects resulting from the generation, management, transboundary movements and disposal of hazardous and other wastes. Note that the United States has not ratified this agreement and, therefore, is not bound to abide by it.
1997	**Kyoto Protocol**	Required the participating 38 developed countries to cut their greenhouse gas emissions back to 5% below 1990 levels. Note that the United States has not ratified this agreement and, therefore, is not bound to abide by it.

One more chapter in which we'll review laboratories, and you're done with the content review portion of the book! Learn the terms and try the drill to make sure you've grasped all of the content in the chapter.

KEY TERMS

Study these terms and you will be sure to write excellent essays.

Policy
Golden Age of Conservation
sustainability
water rights
EPA
NGO
cap-and-trade policy
green taxes
globalization
market permits
Kyoto Protocol

People
Henry David Thoreau & Walden
Rachel Carson & Silent Spring
John Muir & The Sierra Club

Acts
National Environmental Policy Act
International Environmental Protection Act
Pollution Prevention Act
Environmental Education Act

Chapter 10 Drill

Directions: Each of the questions or incomplete statements below is followed by five suggested answers or completions. Select the one that is best in each case. For answers and explanations, see Chapter 13.

1. The release of CFCs was banned under an international treaty written in which of the following cities?

    (A) Montreal
    (B) New York
    (C) New Delhi
    (D) Kyoto
    (E) Washington

2. Which of the following is a function of the Hazardous Materials Transportation Act?

    (A) Dictates how sanitary landfills are to be built
    (B) Defines which materials are labeled "Hazmat"
    (C) Limits point source water pollution
    (D) Defines the maximum amounts of six different air pollutants
    (E) States how mines must be reclaimed

3. Which of the following policies prevents the harassment, capture, injury, or killing of all species of whales, dolphins, seals, and sea lions, as well as walruses, manatees, dugongs, sea otters, and polar bears?

    (A) National Fisheries Act
    (B) CITES
    (C) Clean Water Act
    (D) The Marine Mammal Protection Act
    (E) Maritime Safety Act

4. When a large federal project might have a significant impact on the environment, which of the following must be drafted?

    (A) A cost-benefit analysis
    (B) A interagency review
    (C) A report from the geographical information system
    (D) An environmental impact statement
    (E) A needs statement

5. Rachel Carson described which of the following problems to Americans?

    (A) Pesticide bioaccumulation and poisoning
    (B) The beauty of nature
    (C) Loss of sustainable forest practices
    (D) The problem of the *Tragedy of the Commons*
    (E) America's diminishing oil reserves

6. Which of the following established the "emissions trading policy"?

    (A) Clean Water Act
    (B) Resource Conservation and Recovery Act
    (C) Asbestos Hazard Emergency Response Act
    (D) The Kyoto Protocol
    (E) Clean Air Act

7. All of the following deal with the study of environmental economics EXCEPT

    (A) the study of the impact of goods on the ecosystem
    (B) understanding the economic cost of pollution
    (C) developing policy alternatives to pollution-based industry
    (D) the relative environmental costs and benefits of proposed policies
    (E) the location of new mineral reserves

8. The listing of threatened species and the purchase of land to protect their habitats is legislated in which of the following?

    (A) Federal Noxious Weed Act
    (B) Endangered Species Act
    (C) Convention on Biological Diversity
    (D) Fish and Wildlife Conservation Act
    (E) Migratory Bird Conservation Act

9. Which of the following philosophies would be held by someone with a holistic viewpoint on the management of the earth's resources?

    (A) There are no problems; let's do whatever we want.
    (B) The free market works best, and the government should not interfere.
    (C) We can manage most problems with technology.
    (D) The biodiversity of the earth is the most important issue and we need it to sustain us.
    (E) Humans are the most dominant species on the planet, and we can solve any environmental problem.

10. Which of the following best describes the goal of environmentally sustainable economic growth?

(A) Allowing rapid population growth so there will be more workers
(B) Exploration to find more natural resources
(C) Increasing the quality of goods without depleting the natural resources needed to make the goods
(D) Cutting down forests and replacing them with rangeland
(E) Growing more crops by using larger amounts of fertilizer

11. A cap-and-trade policy might be effective in controlling which type of the following pollutants?

(A) thermal pollution in rivers
(B) organic waste pollution in oceans
(C) underground water pollutants
(D) noise pollution in a city
(E) carbon dioxide in the atmosphere

12. Which of the following can be classified as an NGO?

(A) World Wildlife Fund
(B) Bureau of Land Management
(C) Fish and Wildlife Commission
(D) International Trade Commission
(E) Department of Defense

Free-Response Question

1. The town council of Hilltop Valley has just received a proposal from its advisory panel to build a new coal-fired power plant on the western edge of the county where Hilltop Valley is located. The plant will be located along the back of the county's major river so that water can easily be transported to the plant.

   (a) Describe two parts of the Clean Air Act that will impact the building of this plant. Give one example of a compound that would be impacted under the National Ambient Air Quality Standards (NAAQS) and discuss the regulation of hazardous air pollutant levels.

   (b) Describe two positive and two negative impacts that the plant might have on the region.

   (c) Opponents of the plant say that they will use the Endangered Species Act (ESA) to prevent the plant's construction. Describe how the ESA could be successfully used to stop construction.

   (d) Describe a goal of the Pollution Prevention Act of 1990 and give one example of how the plant can reduce a pollutant it might emit.

REFLECT ACTIVITY

Respond to the following questions:

- For which content topics discussed in this chapter do you feel you have achieved sufficient mastery to answer multiple-choice questions correctly?

- For which content topics discussed in this chapter do you feel you have achieved sufficient mastery to identify hot-button topics and successfully articulate them in free-response questions?

- For which content topics discussed in this chapter do you feel you need more work before you can answer multiple-choice questions correctly?

- For which content topics discussed in this chapter do you feel you need more work before you can successfully identify and articulate them in free-response questions?

- What parts of this chapter are you going to re-review?

- Will you seek further help, outside of this book (such as a teacher, tutor, or AP Central), on any of the content in this chapter—and, if so, on what content?

Chapter 11
AP Environmental Science in the Lab

The College Board has made a conscious decision to define the AP Environmental Science course as having a laboratory component. This means that laboratories you've performed throughout the year in your course will be tested in some way on this exam.

On the other hand, the College Board also decided to be incredibly flexible about the particular labs that should be covered in the course. The variety of different labs that can be performed gives students opportunities to explore various topics in environmental science; this also allows instructors the freedom to adapt their course laboratories to their geographic locations, which can make the course more interesting for students.

In other words, you will be expected to be familiar with the scientific method.

There are several skills that you will be expected to have gained through participating in laboratories in your course, and the College Board describes these skills in the following way. You should be able to

- make your own observations

- research known data (including library research)

- make a hypothesis

- formulate a prediction

- design an experiment

- gather data and formulate results

- draw conclusions

Although scientists work in many different ways, and often one experiment turns into another without the formal recording of a hypothesis or results, you should understand that the scientific method was created in order to provide a very basic outline for scientific investigation.

During your AP Environmental Science course, you have probably been in the lab a number of times and completed many different types of laboratories. You've probably written lab reports and handed them in to your teacher, and you may have had to do a class presentation or two. As we discussed above, the lab experiences you had in your AP Environmental Science course are probably wildly different from the lab experiences that other students have had at different schools—and this is only partly due to the fact that the College Board didn't give your teacher a list of required labs.

On the AP Environmental Science Exam, you could be asked a number of different types of questions in which the lab component from your class will be helpful. However, you won't be able to get full credit simply by remembering the details of what you did in a particular lab. You will have to use your critical thinking skills to answer these questions.

LET'S TRY ONE—A SAMPLE FREE-RESPONSE QUESTION INVOLVING A LAB

For example, a free-response question on the exam might look a lot like the one below.

2. The county range management office has just received a federal grant to study the breeding success of hawks in your area and you—their student intern for the summer—have been chosen to design and perform the research. The range office wants to know how the hawk population at the state prairie reserve compares with the hawk population on the federal grazing land and the private ranch land. The state prairie reserve allows no grazing, cutting, or control burns. The antelope herd that grazed there until a year ago had to be destroyed because of an illness in the herd. Any lightning strikes are extinguished as quickly as possible. The federal grazing lands are rented out to local ranchers whose private lands border the federal land. The management of these lands allows for a specific number of grazing days per year from the mid-spring until the mid-summer—but no cutting, even of damaged trees, and accidental burns are extinguished promptly. Private lands, of course, can be managed as desired.

    (a) Identify and describe what factor or factors will enable you to judge the hawks' breeding success.

    (b) Design a plan of action for determining the hawks' breeding success in the three areas. What will be your control?

    (c) Describe how the land management plans of each of the three areas would affect the success of the hawk population.

    (d) Discuss why a predator such as the hawk would provide evidence of a successful land management plan.

You may be saying to yourself that you know nothing about hawks or the prairie other than what you learned from watching *Little House on the Prairie* reruns. But before you decide that you can't answer this question, think about what you do know.

Take a minute to reread the passage and then the questions. Look at part (a). How do you think you could determine the success of a bird population—or any population, for that matter? Well, one surefire way to assess the health of a population is by seeing how many offspring they produce, and how many of these survive to reproductive age. So, for hawks, the number of offspring produced could be determined by counting the number of eggs laid; the number of offspring that survive to reproductive age could be approximated by counting how many fledgling leave the nest.

So, in answering this part of the question you could cite these two methods, including any others you can think of. However, make sure you don't get over-zealous—don't list too many possible sources of data because after all, you are just one person, you're there for one summer, and the prairie is a big place!

In part (b), you're asked to design a plan of action. You know what your dependent variable is—the success of the hawk population—and you know how you are going to judge this success—by the number of eggs laid and the number of fledglings that leave the nest. The reasoning behind this proposed line of action is that the more eggs laid by individuals, the greater the health of the population. Rates of egg-laying also provide information about the health of the females of the population; specifically, about whether or not they are receiving adequate nutrition. However, keep in mind that most birds of the same species lay about the same number of eggs, but that the difference between clutch sizes may not be an indicator of health—it could be due to chance.

What is the independent variable in this experiment? It's the way in which the land is managed in each of the three areas. Remember that you cannot change the management practices; you have been given those.

What is the control in the experiment? This is a little less clear-cut. You're given three management plans, none of which operates on land that's left untouched. Therefore, you could set up a control at one of the reserves, but it might be difficult to attract hawks and start with an adequate base control number. Another option is to do a **library control.** In a library control, you could determine the maximum number of eggs and fledglings that could be expected based on data collected in previous years. Either one of these types of controls would be acceptable. Let's summarize the plan of action that you might propose.

You will grid off three random plots, each ten acres in area, in each of the three management areas. You will then observe the hawks in these plots and determine the locations of their nests. You will examine each nest (using binoculars because the nests are usually high in trees) and determine the number of eggs in each clutch. You will continue to make observations throughout the summer to determine how many birds are hatched and how many live to the fledgling stage and leave the nest. You will statistically determine the percent success rate of each nest and the overall area to determine the success of the hawks. Then you will compare your data from the three areas.

So, the plan of action that you write in your exam booklet for answering the first parts of the question might include the following points:

Plan of Action

- Using library resources, determine the maximum success rate of offspring production in hawks (control).

- Plot out three ten-acre, random plots on each of the three management areas.

- Survey plots for hawks and nests within each grid and record the locations of the nests.

- Count the number of eggs in each nest.

- Observe the nests and hatchlings, and count the number of hatched birds. Band baby birds for tracking purposes.

- Count the number of fledglings that survive and leave the nest.

- Perform statistical analysis to determine the success of the hawks on each of the three management areas.

The plan of action is your longest answer in this section, but hopefully you can see that it wasn't too difficult. You took the information you learned this year and applied it to the situation, but you didn't need to know much about hawks.

Unfortunately, however, the next parts will require you to know a little more, rather than just asking you to use your imagination. For example, part (c) asks you to describe how the land management plan of each of the three areas would affect the success of the hawk population.

Let's go through them one by one. The state land would be heavy in undergrowth and might have damaged trees located on it. The diversity of the plant population in this area would be relatively low because the grass would predominate—it would be able to grow unchecked. New grass would grow, but probably not many other plant species. On the other hand, the federal land is grazed, sod in this area there will be abundant growth of new grass. Also, due to this natural thinning of the grass, there will probably be some diversity in the vegetation, but without the fall grazing to clean up the dying grass and continue to thin it, it might not be as diverse as the fall grazed land. The private land can be grazed at any time. Controlled fires can also be used on this land to remove dead grass, add nutrients back into the soil, and remove some of the plant species that compete with grass, such as cactus or sage brush. For these reasons, the private land may have more diversity than any of the other land management areas.

You could also theorize that the private land could be overgrazed and thus exhibit low diversity. Really, you could make either presumption and be counted as correct. However, since you are given data on the fire practices and grazing, the first choice might be better. Why is the vegetation diversity important? The more vegetation diversity there is, the more choices for the herbivores in food and habitats—and probably the greater the success of the herbivore. In any event, the more successful the plants are, the more successful the herbivores in the area will be. Since hawks are carnivores, then the more successful the herbivores, the more successful the hawks. See how this line of reasoning naturally led to your answer to part (c)?

You're almost done; let's think about part (d). The success of the management plan would be demonstrated by the success of the hawk because the hawk is the tertiary consumer in the food chain. It cannot be successful if the rest of the food chain isn't strong.

You're done, and you really did not need to know much about the prairie or a hawk! What you did need to know was how to set up a field experiment; the concept of the food chain; and something about land management plans. Even the management plans could have been deduced with some critical thinking.

SOME COMMON AP ENVIRONMENTAL SCIENCE LABS

Below we've listed some of the common labs performed during an AP Environmental Science course, a summary of the procedures you might follow, and the take-home message of each.

Remember that each AP Environmental Science course is different; you may have performed some of these labs, but you have definitely not completed all of them. It's a good idea to review all of these labs and understand their basic workings as well as their intent.

Soil Analysis Lab

- Soil Testing Laboratory: In this lab soil is tested for physical traits and chemical properties, which provide information about the soil's condition and suitability for crops, septic fields, or other purposes. All of the factors tested are listed below.

Chemical properties

1. pH—Clay soil requires more lime (calcium oxide) or alum (aluminum sulfate) to alter its pH than do sandy or loam soils. Iron necessary for plant growth is unavailable when the soil becomes alkaline. Gymnosperms (pine, fir, etc.) grow better in mildly acidic soil.

2. Nitrogen—common plant nutrient component.

3. Phosphorus—common plant nutrient component.

4. Potash—common name for a compound that contains one of the potassium oxides.

Physical characteristics

5. Soil type—sand, silt, clay—use mesh screen, cheese cloth, soil settling in water tubes to determine the percent of each type of particle in the sample.

6. Water-holding capacity—because of the small pores between clay particles, water moves very slowly through clay. Therefore, clay has a greater holding capacity than silt or sand.

7. Permeability—the movement of gas or liquid through the soil.

8. Friability—good soil is rich, light, and easily worked with fingers—this is good for plant growth because roots can easily grow through it.

9. Percent Humus—a measure of soluble organic constituents of soil—the higher the number, the better. Organic soil has qualities of both sand and clay. The small particles of organic soil come together to form larger clusters. Water can be retained inside a cluster, but can move between clusters to percolate. Organic material is also high in nutrients.

10. Buffering capacity of three different types of bedrock such as marble, granite, and basalt when exposed to acid. Marble has high calcium content and is a better buffer than other rocks.

Water Analysis Lab

- Water Chemical and Physical Analysis: Water can be tested from many different sources. Sample kits have tools for many different tests—for example, the LaMotte kit and the spectrophotometer type kit. Below are some commonly performed tests and some of the expected results. It is probably not necessary for you to memorize all the standards, but you should be familiar with them.

1. pH—normal pH of water is between 6.5–8.5.

2. DO—measure of dissolved oxygen in water. Warm-water fish require a minimum of 4 ppm and cold-water fish require 5 ppm.

3. Turbidity—measurement of water clarity. Higher turbidity means there will be low clarity and little sunlight will be able to penetrate the water. A Secchi Disk may be used to measure turbidity, but a more accurate measure can be made with a turbidity unit.

4. Phosphate—an important plant nutrient, typically found in fertilizer and runoff from agricultural lands. Too much leads to eutrophication of water, high BOD and low DO levels. Levels should not exceed 0.025 mg/L in still water and 0.05 mg/L in flowing water.

5. Alkalinity—measure of compounds that shift the pH towards the alkaline. There are no EPA standards, but normal is between 100–250 ppm.

6. BOD—biological oxygen demand, which is required for the aerobic organisms in a body of water. Unpolluted natural waters have a concentration of 5 mg/L or less. High nutrient levels or lots of biological material ready for decomposition are associated with high BOD and vice versa.

7. Chlorine—EPA standards dictate that Cl cannot exceed 250 mg/L. NaCl is applied to roads and parking lots and can run off into streams. Other sources of excess Cl are animal waste, potash fertilizer (KCl), and septic tank effluent. Chlorine is associated with limestone deposits but is not common in other soils, rocks, or minerals.

8. Hardness—a measure of salts composed of calcium, magnesium, or iron. Most water testing kits test for $CaCl_2$. Hard water is more than 121 ppm and soft water is less than 20 ppm.

9. Iron—normal range is 0.1–0.5 ppm.

10. Nitrates—as an important plant nutrient, nitrogen is typically found in fertilizer and is a component of runoff from agricultural lands. Too much leads to eutrophocation of water, high BOD and low DO levels. Over 0.10 mg/L is considered elevated and the EPA limit is 10 mg/L.

11. Total solids—weight of the suspended solids and dissolved solids. All natural waters have some suspended solids, but problem solids are sewage, industrial waste, or excess amounts of algae.

12. Total dissolved solids—naturally occurring in water, but may cause an objectionable taste in drinking water. They are also unsuitable for irrigation because they leave a salt residue on the soil. EPA standard is 500 mg/L, but dissolved solids may range from 20–2,000 mg/L.

13. Fecal Coliform—any bacteria that ferments lactose and produces gas when grown in lactose broth. New tests for this are performed by adding a water sample to a specialized media and observing color changes. Drinking water should show no colonies of growth from the water sample.

Air Quality Labs

- **Air Quality**: Air quality can be assessed using various methods.

- **Particulates**: Sticky paper can be used to collect air particulates from various sources, and then the paper can be examined under a microscope. It is not possible to see the smallest particulates, but they do color the white paper.

- **Ozone**: In this lab, an ecobadge or a homemade potassium iodide gel sampler is hung or worn in order to collect data on tropospheric ozone. The badge or KI sample changes color in the presence of ozone and becomes more intensely colored as the amount of ozone increases.

- **Carbon dioxide**: In this lab, a commercial sampling device is used to determine the amount of carbon dioxide in an air sample. Car exhaust, burning tobacco, or other pollutants can also be sampled.

Other Labs

- **Lichen**: A lichen survey can be used to judge air quality. Lichens are sensitive to air pollution, particularly sulfur dioxide. The most sensitive lichens are the fruticose types, followed by the foliose, and then the crusteous.

- **Scrubber model**: You may have constructed a model of a scrubber to attempt to remove sulfur contaminates from burning coal. You would have used a calcium compound to try to wash the contaminant from the air column.

- **Biodiversity of Invertebrates**: Insects can be counted in an area and then plotted to assess biodiversity. Traps such as fall traps or sticky traps can be set, and bait such as tuna or sugars can be used. The number of different insects captured could be counted, and this number divided into the total number captured would give an idea of the biodiversity of the area. A taxonomic key can be used to determine the number of species and their taxonomic name. A similar setup can be used to determine the impact of an invasive species. In this process, native bugs can be trapped and set up in terrariums; the invasive species is then introduced and the effects documented.

- **Field Trips**: Field trips can be taken to various areas of interest, such as power plants, landfills, or municipal waste treatment plants. The possibilities are almost endless. Think about what you learned from these experiences and how they can help you on this exam.

- **Energy audit**: In this lab, students are asked to use their homes as a laboratory and perform an energy audit, examining the amount of electricity used by their families over a set period of time and then using appliance standards to determine which is the largest energy consumer. A result of this lab is that students suggest how their family's electrical energy needs could be reduced.

- **Food chain**: In this lab, students observe a natural ecosystem and examine the food chain. They identify the organisms that are producers, primary consumers, and secondary consumers; and determine how many levels make up the food chain, what organisms act as decomposers, and the presence of symbiotic relationships.

- **Model Building**: In this lab, models are built to model land formations, coastlines, tectonic plates, mining operations, or any number of other physical formations. In modeling the tectonic plates, students can slide or bump together plate models that are lying on top of a viscous substance representing the magma. From this model, students see how subduction zones and volcanoes work. Other models also allow students to construct a representation of a physical structure in order to better understand that physical structure.

- **Mining**: Students can construct a model of a mine, representing the rock layers and the mineral deposits.

- **LC_{50} (or LD_{50})**: In this lab, students use a kit or a lab procedure to test the concentration (LC) or dose (LD) that would cause the death of 50 percent of a test organism. An example of this procedure is to test the effects of copper (common algaecide and fungicide ingredient) on Daphnia (a tiny species of freshwater crustacea). Daphnia are exposed to a range of Cu levels and then fed fluorescently-tagged sugar. Healthy Daphnia show up blue under UV light.

- **Population Density and Biomass**: This experiment can be performed in a number of different ways. One method is to mark off two plots, then remove all the vegetation from plot one. Plot two is left to grow for an additional period of time. This investigation may also be performed using grass squares in the lab or bottles of algae. The vegetation from plot one is dried and weighed. This provides the baseline data for the amount of biomass present at the beginning of the experiment. After the time period of growth for the second plot, the vegetation is removed from the second plot and treated as the first was. This provides information about the increase in biomass for the experimental time period. Another way this can be expressed is as the productivity of vegetation. By subtracting the initial growth (plot 1) from the experimental growth (plot 2), productivity could be calculated. This would constitute the net productivity for the plot. The grass in the second plot produced X amount of glucose (photosynthate). However, some of this photosynthate must be used to support the needs of the plant, so the number you calculate is the net, not the gross, productivity.

- **Gross and Net Productivity**: If an experiment is performed using bottles of growing algae or duckweed, the gross productivity can also be determined. One of the bottles in this experiment would contain the starting plant material, but instead of being exposed to light, it would be covered with foil to prevent light from entering the bottle. In this bottle, the plants are only able to perform respiration. Therefore, at the end of the experimental time, this bottle should have less biomass than the initial plant material because the plants had to use stored sugar for metabolic process. This would provide you with the metabolic rate information, which, when added to the net productivity, gives you the gross productivity.

- **Other Uses for Vegetation Plots**: Sample plots may be used to examine the biodiversity of vegetation or examine the patterns of plant growth or dispersal. For example, one species of plant can be marked in a plot, and unless a garden was used as the sample plot, the target plant would probably be dispersed in clumps. Remember that random dispersal is not typical in nature, and even distribution typically only occurs in plots planted by humans.

- **Population Growth**: Population growth experiments can involve fast-growing populations such as bacteria, duckweed (*Lemma minor*), rolly polies (sow bugs), or fruit flies (*Drosophila*); and can involve graphing, as does the analysis of human population data. Population growth can be graphed, with time along the *x*-axis and population growth on the *y*-axis. Initially, the curve is a J shape, but it becomes S-shaped. Bacterial growth curves have an extended hook on the J due to the lag time in growth as the bacteria acclimate to the new media.

- **Variables**: In population growth experiments, often other variables are added to samples of bacteria to compare the growth of a normal population to one that has been altered in some way. This can be used to assess the effect of extra nutrients, such as nitrogen or phosphorus, or the negative effects of certain substances on the organism.

- **Turbidity** and **Bacteria**: When studying bacteria, turbidity is commonly observed and recorded. The more turbid (cloudy) the tube, the more growth has taken place. Turbidity can be observed with a spectrophotometer. When the sample tube is inserted into the spectrophotometer, the instrument passes light through the sample tube and measures the amount of light that passed through or was absorbed.

- **Population Size**: In this type of lab, the size of a population of species such as the gypsy moth, caterpillar, or other insect is studied. Remember that a population is a group of individuals of the same species located in a given area. The experiment can involve a collection box and colored stickers that attract caterpillars. The first day, caterpillars are captured and marked. On the second day, the caterpillars are captured and the new and recaptured individuals are marked. By dividing the number of recaptured insects by the size of the population on day one, an estimate of the population that was originally captured can be obtained. On day three, the total number of caterpillars captured is counted. The total number of caterpillars captured on day three is divided by the calculated number from day two—this gives an estimate of the total population.

- **Salinization**: In this lab, students determine if and how salt levels retard seed germination. Students use various dilutions of salts (variety of salt types may be used) to saturate a growing surface. Seeds are sown to determine levels of salt that inhibit germination. This lab mimics what happens in irrigated farm land with salinization.

- **Note**: If you are asked to graph data on this exam, make sure you use reasonable units! Count the graph blocks you're given; this may help you determine an appropriate scale. Also remember that time is usually plotted on the x-axis.

Well, this marks the end of the content review for this exam—congratulations, you've finished! Study the words in the Hit Parade, which follows, and then take the practice exams. If you need extra practice, go through the sample multiple-choice and free-response questions that are offered on the College Board website.

Good luck!

Chapter 12
The AP Environmental Science Hit Parade

ENVIRONMENTAL SCIENCE HIT PARADE

These are all environmental science terms you should know cold before exam day, so make flashcards, cut these out, study them in your sleep—do whatever you have to do to commit them to memory before exam day!

Chapter 4: Earth's Interdependent Systems

abiotic—related to factors or things that are separate and independent from living things; nonliving.

acid—any compound that releases hydrogen ions when dissolved in water. Also, a water solution that contains a surplus of hydrogen ions.

air mass—enormous bodies of air that move as a unit.

A horizon—a soil horizon; the layer below the O horizon is called the A horizon. The A horizon is formed of weathered rock, with some organic material; often referred to as topsoil.

alkaline—a basic substance; chemically, a substance that absorbs hydrogen ions or releases hydroxide ions; in reference to natural water, a measure of the base content of the water.

aquifer—an underground layer of porous rock, sand, or other material that allows the movement of water between layers of nonporous rock or clay. Aquifers are frequently tapped for wells.

arable—land that's fit to be cultivated.

asthenosphere—the part of the mantle that lies just below the lithosphere.

atmosphere—the gaseous mass or envelope surrounding a celestial body—especially the one surrounding the earth which is retained by the celestial body's gravitational field.

barrier island—a long, relatively narrow island running parallel to the mainland, built up by the action of waves and currents and serving to protect the coast from erosion by surf and tidal surges.

biological weathering—any weathering that's caused by the activities of living organisms.

biotic—living or derived from living things.

B horizon—a soil horizon; B receives the minerals and organic materials that are leached out of the A horizon.

chemical weathering—the result of chemical interaction with the bedrock that is typical of the action of both water and atmospheric gases.

C horizon—a soil horizon; horizon C is made up of larger pieces of rock that have not undergone much weathering.

clay—the finest soil, made up of particles that are less than 0.002 mm in diameter.

climate—weather conditions, especially temperature and precipitation, that remain constant over 30 years or more.

conduction—the transmission or conveying of something through a medium or passage, especially the transmission of electric charge or heat through a conducting medium without perceptible motion of the medium itself.

convection—the vertical movement of a mass of matter because of heating and cooling; this can happen in both the atmosphere and Earth's mantle.

convection currents—air currents caused by the vertical movement of air due to atmospheric heating and cooling.

convergent boundary—a plate boundary where two plates are moving toward each other.

coral reef—an erosion-resistant marine ridge or mound consisting chiefly of compacted coral together with algal material and biochemically deposited magnesium and calcium carbonates.

Coriolis effect—the observed effect of the Coriolis force, especially the deflection of an object moving above the earth, rightward in the Northern Hemisphere, and leftward in the Southern Hemisphere, as away from the equator.

crop rotation—the practice of alternating the crops grown on a piece of land to replenish soil nutrients—for example, corn one year, legumes for two years, and then back to corn.

delta—a usually triangular alluvial deposit at the mouth of a river.

divergent boundary—a plate boundary at which plates are moving away from each other. This causes an upwelling of magma from the mantle to cool and form new crust.

doldrums—a region of the ocean near the equator, characterized by calms, light winds, or squalls.

drip irrigation—a method of supplying irrigation water through tubes that literally drip water onto the soil at the base of each plant.

earthquake—the result of vibrations that release energy from within the earth. They often occur as two plates slide past one another at a transform boundary.

El Niño—a climate variation that takes place in the tropical Pacific about every three to seven years, for a duration of about one year.

erosion—the process of soil particles being carried away by wind or water. Erosion moves the smaller particles first and hence degrades the soil to a coarser, sandier, stonier texture.

estuary—the part of the wide lower course of a river where its current is met by the tides.

fault—the place where two tectonic plates abut each other.

Green Revolution—the time after the Industrial Revolution when farming became mechanized and crop yields in industrialized nations boomed as farmers began using large amounts of chemical fertilizers and pesticides.

greenhouse effect—the phenomenon whereby the earth's atmosphere traps solar radiation, caused by the presence in the atmosphere of gases such as carbon dioxide, water vapor, and methane that allow incoming sunlight to pass through but absorb heat radiated back from the earth's surface.

Hadley cell—a system of vertical and horizontal air circulation that creates major weather patterns, predominately in tropical and subtropical regions.

headwaters—the water from which a river rises; a source.

horizon—a layer of soil.

humus—the dark, crumbly, nutrient-rich material that results from the decomposition of organic material, which is also a product of composting organic waste.

hurricane (typhoon, cyclone)—a severe tropical storm originating in the equatorial regions of the Atlantic Ocean or Caribbean Sea or eastern regions of the Pacific Ocean, that travels north, northwest, or northeast from its point of origin, and usually involves high speed winds and heavy rains.

inner core—the molten core of the earth.

insolation—the delivery rate of solar radiation per unit of horizontal surface.

jet stream—a high-speed, meandering wind current, generally moving from a westerly direction at speeds often exceeding 400 km (250 miles) per hour at altitudes of 15 to 25 km (10 to 15 miles).

land degradation—deterioration of land quality (topsoil, organisms, vegetation, water quality), usually caused by its exploitation.

La Niña—a cooling of the ocean surface off the western coast of South America, occurring periodically every 4 to 12 years and affecting Pacific and other weather patterns.

lithosphere—the outer part of the earth, consisting of the crust and upper mantle, approximately 100 km (62 miles) thick.

loamy—soil composed of a mixture of sand, clay, silt, and organic matter.

mantle—the layer of the earth between the crust and the core.

monoculture—the cultivation of a single crop on a farm or in a region or country; a single, homogeneous culture without diversity or dissension.

monsoon—a wind system that influences large climatic regions and reverses direction seasonally.

O horizon—the uppermost horizon of soil. It is primarily made up of organic material, including waste from organisms, the bodies of decomposing organisms, and live organisms.

physical (mechanical) weathering—any process that breaks rock down into smaller pieces without changing the chemistry of the rock; typically wind and water.

plate boundaries—the edges of tectonic plates.

prior appropriation—when water rights are given to those who have historically used the water in a certain area.

rain shadow effect—the low-rainfall region that exists on the leeward (downwind) side of a mountain range. This rain shadow effect is the result of the mountain range's causing precipitation on the windward side.

red tide—a bloom of dinoflagellates that causes reddish discoloration of coastal ocean waters. Certain dinoflagellates of the genus *Gonyamlax* produce toxins that kill fish and contaminate shellfish.

R horizon—the bedrock, which lies below all of the other layers of soil, is referred to as the R horizon.

riparian right—the right, as to fishing or to the use of a riverbed, of one who owns riparian land (the land adjacent to a river or stream).

salinization—occurs when soil becomes waterlogged from excess irrigation and then dries out. As the water evaporates, the salt crystallizes and forms a layer on the soil surface. This excess of salt prevents the growth of plants.

sand—the coarsest soil, with particles 0.05–2.0 mm in diameter.

silt—soil with particles 0.002–0.05 mm in diameter.

Southern Oscillation—the atmospheric pressure conditions corresponding to the periodic warming of El Niño and cooling of La Niña.

subduction zone—in tectonic plates, the site at which an oceanic plate is sliding under a continental plate.

thermocline—a layer in a large body of water, such as a lake, that sharply separates regions differing in temperature, so that the temperature gradient across the layer is abrupt.

thermosphere—the outermost shell of the atmosphere, between the mesosphere and outer space, where temperatures increase steadily with altitude.

topsoil—the A horizon of soil is often referred to as topsoil and is most important for plant growth.

trade winds—the more or less constant winds blowing in horizontal directions over the earth's surface, as part of Hadley cells.

transform boundary—also known as transform faults, boundaries at which plates are moving past each other, sideways.

tropical storm—a cyclonic storm having winds ranging from approximately 48 to 121 km (30 to 75 miles) per hour.

upwelling—a process in which cold, often nutrient-rich, waters from the ocean depths rise to the surface.

volcanoes—an opening in the earth's crust through which molten lava, ash, and gases are ejected.

watershed—the region draining into a river system or other body of water.

water-scarce—countries that have a renewable annual water supply of less than 1,000 m^3 per person.

water-stressed—countries that have a renewable annual water supply of about 1,000–2,000 m^3 per person.

weather—the day-to-day variations in temperature, air pressure, wind, humidity, and precipitation mediated by the atmosphere in a given region.

weathering—the gradual breakdown of rock into smaller and smaller particles, caused by natural chemical, physical, and biological factors.

wetlands—a lowland area, such as a marsh or swamp, that is saturated with moisture, especially when regarded as the natural habitat of wildlife.

Chapter 5: The Inhabitants of Planet Earth and Their Relationships

ammonification—the production of ammonia or ammonium compounds in the decomposition of organic matter, especially through the action of bacteria.

assimilation—the process in which plants absorb ammonium (NH_3), ammonia ions (NH_4^+), and nitrate ions (NO_3) through their roots.

autotroph—producers; organisms that can produce their own organic compounds from inorganic compounds. They use energy from the sun or from the oxidation of inorganic substances.

bioaccumulation—the accumulation of a substance, such as a toxic chemical, in various tissues of a living organism.

biological extinction—true extermination of a species. There are no individuals of this species left on the planet.

biomagnification—the process by which the concentration of toxic substances increases in each successive link in the food chain.

biosphere—the part of the earth and its atmosphere where living organisms exist or that is capable of supporting life.

carnivore—an animal that only consumes other animals.

chemotroph (chemoautotroph)—an organism, such as a bacterium or protozoan, that obtains its nourishment through the oxidation of inorganic chemical compounds, as opposed to photosynthesis.

climax community—a stable, mature community in a successive series that has reached equilibrium after having evolved through stages and adapted to its environment.

combustion—the process of burning.

commercial or economic extinction—a few individuals exist but the effort needed to locate and harvest them is not worth the expense.

community—formed from populations of different species occupying the same geographic area.

competitive exclusion—the process that occurs when two different species in a region compete and the better adapted species wins.

consumer—an organism that must obtain food energy from secondary sources, for example, by eating plant or animal matter.

decomposer—bacteria or fungi that absorb nutrients from nonliving organic matter like plant material, the wastes of living organisms, and corpses. They convert these materials into inorganic forms.

denitrification—the process by which specialized bacteria (mostly anaerobic bacteria) convert ammonia to NO_3, NO_2, and N_2 and release it back to the atmosphere.

detritivore—organisms that derive energy from consuming nonliving organic matter, such as dead animals or fallen leaves. Earthworms and many species of fungi are detritivores.

ecological extinction—there are so few individuals of a species that this species can no longer perform its ecological function.

ecological succession—transition in species composition of a biological community, often following ecological disturbance of the community; the establishment of a biological community in any area virtually barren of life.

edge effect—the condition in which, at ecosystem boundaries, there is greater species diversity and biological density than there is in the heart of ecological communities.

energy pyramid—the structure obtained if we organize the amount of energy contained in producers and consumers in an ecosystem by kilocalories per square meter, from largest to smallest.

evaporation—to convert or change into a vapor.

evolution—change in the genetic composition of a population during successive generations as a result of natural selection acting on the genetic variation among individuals and resulting in the development of new species.

extinction—the death of an entire species; permanent inactivity.

food chain—a succession of organisms in an ecological community that constitutes a continuation of food energy from one organism to another as each consumes a lower member and, in turn, is preyed upon by a higher member.

food web—a complex of interrelated food chains in an ecological community.

Gause's principle—states that no two species can occupy the same niche at the same time, and that the species that is less fit to live in the environment will either relocate, die out, or occupy a smaller niche.

Gross Primary Productivity—the amount of sugar that the plants produce in photosynthesis, and subtracting from it the amount of energy the plants need for growth, maintenance, repair, and reproduction.

habitat—the area or environment where an organism or ecological community normally lives or occurs.

habitat fragmentation—when the size of an organism's natural habitat is reduced, or when development occurs that isolates a habitat.

heterotroph—an organism that cannot synthesize its own food and is dependent on complex organic substances for nutrition.

indigenous species—species that originate and live, or occur naturally, in an area or environment.

invasive species—an introduced, nonnative species.

keystone species—a species whose very presence contributes to an ecosystem's diversity and whose extinction would consequently lead to the extinction of other forms of life.

Law of Conservation of Matter—states that matter can neither be created nor destroyed.

mutualism—a symbiotic relationship in which both species benefit.

natural selection—the process by which, according to Darwin's theory of evolution, only the organisms best adapted to their environment tend to survive and transmit their genetic characteristics in increasing numbers to succeeding generations, while those less adapted tend to be eliminated.

Net Primary Productivity (NPP)—the amount of energy that plants pass on to the community of herbivores in an ecosystem.

niche—the total sum of a species' use of the biotic and abiotic resources in its environment.

nitrification—the process in which soil bacteria convert ammonium (NH_4^+) to a form that can be used by plants; nitrate, or NO_3.

nitrogen fixation—the conversion of atmospheric nitrogen into compounds, such as ammonia, by natural agencies or various industrial processes.

omnivores—organisms that consume both producers and primary consumers.

parasitism—a symbiotic relationship in which one member is helped by the association and the other is harmed.

photosynthesis—the process in green plants and certain other organisms by which carbohydrates are synthesized from carbon dioxide and water using light as an energy source. Most forms of photosynthesis release oxygen as a byproduct.

pioneer species—organisms in the first stages of succession.

population—a group of organisms of the same species that live in the same area.

predation—when one species feeds on another.

primary consumers—this category includes organisms that consume producers (plants and algae).

primary succession—when ecological succession begins in a virtually lifeless area, such as the area behind a moving glacier.

producer—an organism that is capable of converting radiant energy or chemical energy into carbohydrates.

realized niche—when a species occupies a smaller niche than it would in the absence of competition.

reservoir—a place where a large quantity of a resource sits for a long period of time.

respiration—the process in which animals (and plants!) breathe and give off carbon dioxide from cellular metabolism.

residency time—the amount of time a resource spends in a reservoir or an exchange pool.

secondary consumers—organisms that consume primary consumers.

species—organisms that are capable of breeding with one another and incapable of breeding with other species.

symbiotic relationships—close, prolonged associations between two or more different organisms of different species that may, but do not necessarily, benefit the members.

tertiary consumers—organisms that consume secondary consumers or other tertiary consumers.

transpiration—the act or process of transpiring, or releasing water vapor, especially through the stomata of plant tissue or the pores of the skin.

trophic level—each of the feeding levels in a food chain.

Chapter 6: Population Ecology

age-structure pyramids—graphical representations of populations' ages.

albedo—the fraction of solar energy that is reflected back into space.

biotic potential—the amount that the population would grow if there were unlimited resources in its environment.

birth rate (crude birth rate)—the number of live births per 1,000 members of the population in a year.

carrying capacity—the maximum population size that can be supported by the available resources in a region.

death rate (crude death rate)—is equal to the number of deaths per 1,000 members of the population in a year.

demographic transition model—a model that's used to predict population trends based on the birth and death rates as well as economic status of a population.

ecological footprint—the amount of the earth's surface that's required to supply the needs of and dispose of the waste from a particular population.

emigration—the movement of individuals out of a population.

genetic drift—the random fluctuations in the frequency of the appearance of a gene in a small isolated population, presumably owing to chance, rather than natural selection.

immigration—the movement of individuals into a population.

***k*-selected**—organisms that reproduce later in life, produce fewer offspring, and devote significant time and energy to the nurturing of their offspring.

logistic population growth—when a population is well below the size dictated by the carrying capacity of its region, it will grow exponentially, but as it approaches the carrying capacity, its growth rate will decrease and the size of the population will eventually become stable.

population density—the number of individuals of a population that inhabit a certain unit of land or water area.

replacement birth rate—the number of children a couple must have in order to replace themselves in a population.

***r*-selected**—organisms that reproduce early in life and often and have a high capacity for reproductive growth.

total fertility rate—the number of children an average woman will bear during her lifetime; this information is based on an analysis of data from preceding years in the population in question.

Chapter 7: Resource Utilization

agroforestry—when trees and crops are planted together, creating a mutualistic symbiotic relationship between them.

aquaculture—the raising of fish and other aquatic species in captivity for harvest.

bottom trawling—a fishing technique in which the ocean floor is literally scraped by heavy nets that smash everything in their path.

by-catch—any other species of fish, mammals, or birds that are caught that are not the target organism.

capture fisheries—fish production in which fish are caught in the wild and not raised in captivity for consumption.

clear-cutting—the removal of all of the trees in an area.

conservation—the management or regulation of a resource so that its use does not exceed the capacity of the resource to regenerate itself.

consumption—the day-to-day use of environmental resources such as food, clothing, and housing.

contour plowing—a process in which rows of crops are plowed across the hillside; this prevents the erosion that can occur when rows are cut up and down on a slope.

deforestation—the removal of trees for agricultural purposes or purposes of exportation.

driftnets—nets that drift free in the water and indiscriminately catch everything in their path.

ecosystem capital—the value of natural resources.

fishery—the industry or occupation devoted to the catching, processing, or selling of fish, shellfish, or other aquatic animals.

greenbelt—open or forested areas built at the outer edge of a city.

ground fires—smoldering fires that take place in bogs or swamps and can burn underground for days or weeks. Originating from surface fires, ground fires are difficult to detect and extinguish.

intercropping—(also called strip cropping) is the practice of planting bands of different crops across a hillside.

long lining—in fishing, the use of long lines that have baited hooks and will be taken by numerous aquatic organisms.

malnutrition—poor nutrition that results from an insufficient or poorly balanced diet.

mineral deposit—an area where a particular mineral is concentrated.

mining—the excavation of the earth for the purpose of extracting ore or minerals.

monoculture—when just one type of plant is planted in a large area.

natural resources—biotic and abiotic natural ecosystems.

nonrenewable resources—resources that are often formed by very slow geologic processes, so we consider them incapable of being regenerated within the realm of human existence.

no-till methods—refers to when farmers plant seeds without using a plow to turn the soil.

old growth forest—one that has never been cut; these forests have not been seriously disturbed for several hundred years.

overgrazed—when grass is consumed by animals at a faster rate than it can regrow.

preservation—the maintenance of a species or ecosystem in order to ensure its perpetuation, with no concern as to their potential monetary value.

production—the use of environmental resources for profit.

renewable resources—refers to resources, such as plants and animals, which can be regenerated if harvested at sustainable yields.

second growth forests—areas where cutting has occurred and a new, younger forest has arisen.

selective cutting—the removal of select trees in an area; this leaves the majority of the habitat in place and has less of an impact on the ecosystem.

shelter-wood cutting—when mature trees are cut over a period of time (usually 10–20 years); this leaves mature trees, which can reseed the forest, in place.

silviculture—the management of forest plantations for the purpose of harvesting timber.

slash and burn—when an area of vegetation is cut down and burned before being planted with crops.

surface fires—fires that typically burn only the forest's underbrush and do little damage to mature trees. These fires actually serve to protect the forest from more harmful fires by removing underbrush and dead materials that would burn quickly and at high temperatures.

tailings—piles of gangue, which is the waste material that results from mining.

traditional subsistence agriculture—when each family in a community grows crops for themselves and relies on animal and human labor to plant and harvest crops.

terracing—creating flat platforms in the hillside that provide a level planting surface, which reduces soil runoff from the slope.

tree farms—also known as plantations, these are planted and managed tracts of trees of the same age that are harvested for commercial use.

uneven-aged management—the broad category under which selective cutting and shelter-wood cutting fall; selective deforestation.

Chapter 8: Energy

active collection—the use of devices, such as solar panels, to collect, focus, transport, or store solar energy.

anthracite—the cleanest-burning coal; almost pure carbon.

barrels—the unit used to describe the volume of fossil fuels.

bituminous—the second-purest form of coal.

crude oil—the form petroleum takes when in the ground.

energy—the capacity to do work.

fission—a nuclear reaction in which an atomic nucleus, especially a heavy nucleus such as an isotope of uranium, splits into fragments, usually two fragments of comparable mass, releasing from 100 million to several hundred million electron volts of energy.

fossil fuel—a hydrocarbon deposit, such as petroleum, coal, or natural gas, derived from living matter of a previous geologic time and used for fuel.

First Law of Thermodynamics—says that energy can neither be created nor destroyed; it can only be transferred and transformed.

fly ash—a waste product produced by the burning of coal.

half-life—the amount of time it takes for half of a radioactive sample to disappear.

Hubbert peak (also known as **peak oil**)—an influential theory that concerns the long-term rate of conventional oil (and other fossil fuel) extraction and depletion. It predicts that future world oil production will soon reach a peak and then rapidly decline.

hydroelectric power—power generated using water.

kinetic energy—the energy of motion.

lignite—the least pure coal.

nuclear fusion—the process of fusing two nuclei.

overburden—the rocks and earth that are removed when strip mining for a commercially valuable mineral resource.

passive solar energy collection—the use of building materials, building placement, and design to passively collect solar energy that can be used to keep a building warm or cool.

peak oil (see **Hubbert peak**)

petroleum—oil, a hydrocarbon that forms as sediments are buried and pressurized.

photovoltaic cell (PV cell)—a semiconductor device that converts the energy of sunlight into electric energy.

potential energy—energy at rest, or stored energy.

proven reserve—an estimate of the amount of fossil fuel that can be obtained from reserve.

radiant energy—sunlight.

scrubbers—devices containing alkaline substances that precipitate out much of the sulfur dioxide from industrial plants air effluent.

Second Law of Thermodynamics—says that the entropy (disorder) of the universe is increasing. One corollary of the Second Law of Thermodynamics is the concept that, in most energy transformations, a significant fraction of energy is lost to the universe as heat.

strip mining—involves the removal of the earth's surface all the way down to the level of the mineral seam.

subbituminous—the third purest form of coal.

underground mining—involves the sinking of shafts to reach underground deposits. In this type of mining, networks of tunnels are dug or blasted and humans enter these tunnels in order to manually retrieve the coal.

wind farm—a group of modern turbines.

Chapter 9: Pollution

acid precipitation—acid rain, acid hail, acid snow; all of which occur as a result of pollution in the atmosphere.

acute effect—the effect caused by a short exposure to a high level of toxin.

catalytic converter—a platinum-coated device that oxidizes most of the VOCs and some of the CO that would otherwise be emitted in exhaust, converting them to CO_2.

closed-loop recycling—when materials, such as plastic or aluminum, are used to rebuild the same product. An example of this is the use of the aluminum from aluminum cans to produce more aluminum cans.

composting—a process that allows the organic material in solid waste to be decomposed and reintroduced into the soil, often as fertilizer.

building-related illness—when the signs and symptoms of an illness can be attributed to a specific infectious organism that resides in the building.

chronic effect—an effect that results from long-term exposure to low levels of toxin.

deep well injection—drilling a hole in the ground that's below the water table to hold waste.

disease—occurs when infection causes a change in the state of health.

dose-response analysis—a process in which an organism is exposed to a toxin at different concentrations, and the dosage that causes the death of the organism is recorded.

dose-response curve—the result of graphing a dose-response analysis.

ED$_{50}$—the point at which 50 percent of the test organisms show a negative effect from a toxin.

global warming—an intensification of the Greenhouse Effect due to the increased presence of heat-trapping gases in the atmosphere.

gray smog (industrial smog)—smog resulting from emissions from industry and other sources of gases produced by the burning of fossil fuels, especially coal.

hazardous waste—any waste that poses a danger to human health; it must be dealt with in a different way from other types of waste.

heat islands—urban areas that heat up more quickly and retain heat better than nonurban areas.

high-level radioactive waste—radioactive wastes that produce high levels of ionizing radiation.

industrial smog—(see **gray smog**)

infection—the result of a pathogen invading a body.

LD$_{50}$—the point at which 50 percent of the test organisms die from a toxin.

leachate—the liquid that percolates to the bottom of a landfill.

low-level radioactive waste—radioactive wastes that produce low levels of ionizing radiation.

noise pollution—any noise that causes stress or has the potential to damage human health.

non-point source pollution—pollution that does not have a specific point of release.

open-loop recycling—when materials are reused to form new products.

ozone holes—the thinning of the ozone layer over Antarctica (and to some extent, over the Arctic).

pathogens—bacteria, virus, or other microorganisms that can cause disease.

photochemical smog—usually formed on hot sunny days when NO_x compounds, VOCs, and ozone combine to form smog with a brownish hue.

point source pollution—a specific location from which pollution is released; an example of a point source location is a factory where wood is being burned.

poison—any substance that has an LD$_{50}$ of 50 mg or less per kg of body weight.

physical treatment—in a sewage treatment plant, the initial filtration that is done to remove debris such as stones, sticks, rags, toys, and other objects that were flushed down the toilet.

primary pollutants—pollutants that are released directly into the lower atmosphere.

primary treatment—when physically treated sewage water is passed into a settling tank, where suspended solids settle out as sludge; chemically treated polymers may be added to help the suspended solids separate and settle out.

risk assessment—calculating risk, or the degree of likelihood that a person will become ill upon exposure to a toxin or pathogen.

risk management—using strategies to reduce the amount of risk (the degree of likelihood that a person will become ill upon exposure to a toxin or pathogen).

secondary pollutants—pollutants that are formed by the combination of primary pollutants in the atmosphere.

secondary treatment—the biological treatment of wastewater in order to continue to remove biodegradable waste.

sick building syndrome—a condition in which the majority of a building's occupants experience certain symptoms that vary with the amount of time spent in the building, without being able to identify a specified cause or illness.

sludge—the solids that remain after the secondary treatment of sewage.

sludge processor—a tank filled with aerobic bacteria that's used to treat sewage.

solid waste—can consist of hazardous waste, industrial solid waste, or municipal waste. Many types of solid waste provide a threat to human health and the environment.

stationary sources—non-moving sources of pollution, such as factories.

Superfund Program—a program funded by the federal government and a trust that's funded by taxes on chemicals; identifies pollutants and cleans up hazardous waste sites.

threshold dose—the dosage level of a toxin at which a negative effect occurs.

toxicity—the degree to which a substance is biologically harmful.

toxin—any substance that is inhaled, ingested, or absorbed at dosages sufficient to damage a living organism.

tropospheric ozone—ozone that exists in the trophosphere.

U.S. Noise Control Act—gave the EPA power to set emission standards for major sources of noise, including transportation, machinery, and construction.

vector—the carrier organism through which pathogens can attack, such as a tick.

wastewater—any water that has been used by humans. This includes human sewage, water drained from showers, tubs, sinks, dishwashers, washing machines, water from industrial processes, and storm water runoff.

Waste-to-Energy (WTE) program—when the energy released from waste incineration is used to generate electricity.

Chapter 10: Culture, Society, and Environmental Quality

green tax—a fiscal policy that lowers taxes on income, including wages and profit, and raises taxes on consumption, particularly the unsustainable consumption of non-renewable resources.

market permits—when companies are allowed to buy permits that allow them a certain amount of discharge of substances into certain environmental outlets. If they can reduce their amount of discharge, they are allowed to sell the remaining portion of their permit to another company.

Chapter 13
Chapter Drills:
Answers and Explanations

CHAPTER 4 DRILL

Multiple-Choice Answers

1. **B** A seismograph measures the magnitude of earthquakes by recording Earth's movement, or seismic waves. We use the Richter Scale to measure the amplitude of the highest S-wave of an earthquake.

2. **B** The aurora borealis occurs in the thermosphere, or ionosphere. This is the highest level of the atmosphere, extending 60 miles and higher. The aurora borealis is caused by electrons from the Sun striking oxygen atoms in the thermosphere.

3. **E** Climate is the 30-year average of temperature and precipitation in a certain area, and the temperature and precipitation of a region are the most important factors in determining the rate of plant growth. Finally, since the amount of plant matter determines the amount of animal life, the communities found in a particular habitat are directly related to its temperature and precipitation.

4. **B** As the Sun's light passes through the atmosphere, it strikes the solid earth. The earth, with its soil, water, buildings, asphalt, and concrete, absorbs this radiant energy. This energy is then radiated back into the atmosphere as infrared radiation. This radiation can be reflected back into the atmosphere (the greenhouse effect) or it can pass back into space.

5. **C** All of the answer choices listed are parts of the hydrosphere except for choice (C), parent rock.

6. **A** Lakes are divided into various zones depending on light and temperature. The littoral zone is where rooted plants live, while the limnetic zone is open, sunlit water that's generally warm. The profundal zone is the deep open water where no photosynthesis occurs and the benthic zone is the cold, dark zone at the bottom. Only organisms that can tolerate cold and low oxygen levels can live in the benthic zone.

7. **D** Approximately 75 percent of the earth's surface is covered by water; this includes both saltwater and freshwater.

8. **D** An area where saltwater and freshwater mix that has a very high level of productivity is correctly called an estuary. Estuaries are sites where the "arm" of the sea extends inland to meet the mouth of a river. Estuaries are often rich with many different types of plant and animal species, because the fresh water in these areas usually has a high concentration of nutrients and sediments.

9. **A** Both freshwater and saltwater bodies experience a seasonal movement of water from the cold and nutrient-rich bottom to the surface; this is called an upwelling. Upwellings are composed of cold, nutrient-rich waters.

10. **B** Unconfined aquifers are ones where water is free to move vertically or horizontally through an area.

11. **D** Deer are herbivores feeding on plants, not members of the decay food web found in the soil.

12. **E** Ozone in the atmosphere is contained in the stratosphere.

13. **B** Most of our weather patterns occur in the troposphere.

14. **D** Convection currents are the vertical heating and cooling of air masses.

15. **C** Wind is the movement of air between masses with different pressures.

16. **A** The atmosphere can be defined as the collection of all gases held to the earth by gravity.

Free-Response Answer

1. First of all, you might need some information from Chapter 4 to answer this question. However, this will give you an opportunity to see how prepared you are for this exam, before you review all of the major topics you'll need to know. How did your answer compare to the one below?

(a) The runoff volume would be much greater in the clear-cut plot than the forested plot. In the forested area the trees help hold water in the soil. Also, the leaves slow down the speed of the raindrops, making them have less of an impact on the soil. The roots also help bind the soil and hold it in place, so there is no erosion. In the cut area the water stays closer to the surface, so there is more water for runoff. The rain can fall onto the soil with full force and erosion can take place. The soil is not held together and the particles have a greater likelihood of moving.
(2 points maximum—1 for the correct answer and 1 point for a correct explanation)

(b) The phosphate levels would be much higher in the runoff from the cut plot. The phosphate would be leached out of the soil in the clear-cut plot because of all the water running off the soil. Another explanation would be that the trees absorb most of the phosphorus out of the soil. There would be less phosphate in the soil of the forested land, so the runoff would contain less.
(2 points maximum—1 for the correct answer and 1 point for a correct explanation)

(c) There are several possible negative effects. First, the added nutrients could cause an algal bloom in the stream. The bloom might make the stream less habitable for fish or insect larvae. As the algae decompose, the amount of dissolved oxygen would go down. The sediment might increase the water's turbidity, making it cloudier, and thus lowering the ability of producers to live in the stream. Also, the increased water volume might cause more erosion, or possibly flooding farther downstream. Finally, the lack of shade would increase the water's temperature. This increase would lower the dissolved oxygen (DO) levels.
(2 points maximum—1 for the correct answer and 1 point for a correct explanation)

(d) Because of the very rapid rate of decay and high metabolism of the living plants, there is little organic material in rain forest soil. Anything that falls to the forest floor is quickly decomposed, and the remains are rapidly absorbed by plants. When the forest is cut down, this soil is directly exposed to the large amounts of rain that falls in these forests. The rain quickly washes away the remaining organic matter, leaving even fewer nutrients in the soil.
(4 points maximum—2 for the correct explanation of why there are few nutrients and 2 for the correct explanation of the rapid loss of the remaining nutrients)

CHAPTER 5 DRILL

Multiple-Choice Answers

1. **C** The relationship between the bird and the tick is best described as parasitism, which is one type of symbiotic relationship. In this case, the parasite is the tick; it benefits from the relationship, while the bird is harmed.

2. **B** This concept is called competitive exclusion. The idea behind competitive exclusion is that two species that share the same niche cannot infinitely exist in the same ecosystem; eventually one will prove to be more fit and out-compete the other.

3. **D** In resource partitioning, different species can use slightly different parts of the habitat to avoid direct competition.

4. **E** One aspect of the roles of both predators and parasites in an ecosystem is that they generally eliminate the weakest members of a population. The weakest individuals are those who are young, sick, or old; these individuals are eliminated, leaving the best-adapted organisms to survive and reproduce.

5. **A** (A) is false. Small adult plants are found in early succession stages. The other characteristics are all true of climax communities.

6. **C** Food chains show feeding relationships. The energy in a food chain flows from the Sun, through producers (plants), to primary consumers, to secondary consumers, and then to tertiary consumers. In other words, energy flows from prey to predator: choice (C).

7. **A** Sulfur is stored as sulfur salts in rocks and as sulfur dioxide in the atmosphere.

8. **E** Microevolution is the process where a population shows small scale genetic changes over a short period of time. Antibiotic resistance in bacteria occurs when overuse of antibiotics kills off the susceptible cells and the resistant ones are able to thrive.

9. **D** Grasslands can support large numbers of animals and the herd helps with protection from predators.

10. **E** When late succession plants are not disturbed by early succession plants, they are exhibiting tolerance.

11. **D** When succession starts from an area where humans once farmed, it's called secondary succession.

12. **C** When a community starts from bare rock, this process is called primary succession.

13. **A** The tendency of an ecosystem to maintain its overall structure is known as inertia.

14. **B** A disturbance is an event that will instigate the process of succession.

15. **C** Native species are those species that normally live and thrive in a particular habitat.

16. **B** Keystone species are species that play a pivotal role in the habitat.

17. **E** Indicator species are species whose decline indicates damage to a habitat.

18. **D** Alien species are not native to a particular environment; they have been introduced to the environment.

19. **A** A specialist species is one that has a narrow niche, and can only live in a certain habitat.

Free-Response Answer

1. **(a)** Compound X can enter the pond from surface water runoff that carries the compound. It could also enter the pond by being carried in by rain, snow, or dust that falls into the water. The substance might get carried in by ground water from a polluted aquifer.
    (2 points maximum, 1 point for the correct name and 1 point for a correct description of the process)

    (b) Water (0.1 ppb) → Zooplankton (0.2 ppb) → Small fish (0.1 ppm) → Predatory fish (1.0 ppm) → Hawk (3.0 ppm)
    (2 points maximum, 1 point for correct order and 1 point for the concentrations)

    (c) Bioaccumulation and biomagnification are two of the most important processes to know for the test. In bioaccumulation, fat-soluble molecules accumulate and stay in the fatty tissues of animals since they can not dissolve in water. Biomagnification occurs when compounds are passed from prey to predator. Since a predator needs to eat a lot of prey, each of the prey organisms gives some of the compound to the predator. The compounds accumulate and the concentration becomes much higher than you would expect to be in the environment.
    (4 points maximum, 2 points for a correct description of each process)

    (d) Mercury and PCB's (polychlorinated biphenyl) are two very common molecules that bioaccumulate and biomagnify. For PCB's—skin conditions such as chloracne and rashes, changes in blood and urine that may indicate liver damage, dermal and ocular lesions, irregular menstrual cycles, lowered immune response, fatigue, headache, cough, and in children poor cognitive development. For mercury—itching, burning or pain, skin discoloration (pink cheeks, fingertips and toes), swelling, desquamation (shedding of skin), sweating, tachycardia, increased salivation, and hypertension. Affected children may show red cheeks, nose and lips, loss of hair, teeth, and nails, transient rashes, muscle weakness, and increased sensitivity to light. Other symptoms may include kidney disfunction, emotional lability, memory impairment, or insomnia.
    (2 points maximum, 1 for the compound and 1 of a correct symptom associated with the compound)

CHAPTER 6 DRILL

Multiple-Choice Answers

1. **C** Remember that with EXCEPT questions, you're looking for the answer that does not fit the statement. If you read through them, you'll see that choice (C) is the only factor that does not measure some characteristic of populations. Habitat is important in determining a population's size, but it is not a way to measure a population.

2. **D** Emigration refers to the movement of individuals that are leaving, or emigrating from, a population. Birth rate is the number of births per thousand; carrying capacity is the maximum number that can live

sustainably in a habitat; immigration is the movement of individuals into a population; and environmental resistance is all of the factors in a habitat that limit a population's growth.

3. **C** To do this problem, remember the rule of 70, which approximates the time it takes for a population to double (called its doubling time). Take the rate of change (2 percent), and divide that number into 70. So, $\frac{70}{2} = 35$ the number of years that it will take this population to double in size.

4. **E** Age-structure pyramids are often constructed using data about the number and gender of various age groups in a population. Generally, the female population is shown on the right side of the pyramid, and the male population is situated on the left side. The length of each age group indicates how the total number of individuals in that age group compares to the other age groups, and the population as a whole.

5. **E** k-selected organisms reproduce later in life, produce fewer offspring, and devote significant time and energy to the nurturing of their offspring. For these species, it is important to preserve as many members of the offspring as possible because they produce so few; parents have a tremendous investment in each individual offspring. Some examples of k-selected species are humans, lions, and cows.

6. **A** Organisms like the hare and the lynx (predator and prey) exhibit regular changes in their population (every 10 years) in a pattern known as boom-and-bust. For the other choices, an irruptive population is very large and then very small; an irregular population behaves in a chaotic manner; a logistic population doubles in a short time; and a stable population varies only slightly above and below its carrying capacity over time.

7. **B** The demographic transition model is used to study countries' transitions from one type of economy to another; specifically, how the transition affects the population. The four stages of the transition are pre-industrial (high birth and death rates), transitional (high birth rates and low death rates), industrial (declining birth and death rates), and finally postindustrial (very low birth and death rates).

8. **C** HIV/AIDS is correct. The other diseases are seen more commonly in Western Europe and North America. HIV/AIDS spreads very rapidly because it is caused by an easily transmittable virus; the other diseases listed are not communicable—they cannot be passed from person to person.

9. **A** One billion is the figure that is closest to the current population in India.

10. **E** (E) is the only answer choice that lists a factor that is density independent. Destruction by humans (or natural event) would occur whether the population density was low or high. Density dependent factors only influence a population when the density is high. You might have thought that this answer choice was incorrect because you reasoned that, as a population increases in size, it could get so large that it would degrade its environment. However, remember carrying capacity! Habitats only tolerate the existence of a certain number of individuals in a population.

11. **C** Environmental resistance factors such as competition, parasites, or a lack of reources are factors that slow a population's growth. Factors might include competition, parasites, or a lack of resources.

12. **A** Births and immigration add individuals to a population whereas deaths and emigration remove individuals. The difference between gain and loss is the growth.

13. **E** Habitat conservation is the only factor that promotes species growth. All other factors cause a population to decline.

14. **C** The number of people who die per 1,000 in the population is known as the mortality rate of the population.

15. **D** The life expectancy of a person in a population is defined as the average number of years a person can be expected to live.

16. **B** The total fertility rate is used to describe the number of children a woman will bear during her lifetime, and this information is based on an analysis of data from preceding years in the population in question. Total fertility rates are predictions that provide a rough estimate, but they can't be depended on because they assume that the conditions of the past will be the conditions of the future.

17. **A** The birth rate of a population is the number of individuals born per 1,000 in the population.

18. **E** The replacement birth rate of a human population refers to the number of children a couple must have in order to replace themselves in a population. While you might automatically think that the answer is always two, in reality it is slightly higher to compensate for the deaths of children, the existence of non-child-bearing females in the population, and other factors.

19. **C** In developing countries experiencing poverty, women tend to have more children than women in developed nations. While this is in part the result of a lack of available birth control, this cultural phenomenon is also the result of the need for these children to go to work and provide an economic "safety net" for families. Also, at poverty level, infant mortality rates are higher, which means that women must have more children to even attain replacement rate.

Free-Response Answer

1. **(a)** The carrying capacity is the maximum number of individuals that a habitat can sustain for a long period of time. If a population exceeds the carrying capacity, there will be a die-off of individuals until the population dips below the carrying capacity. When the population is lower than the carrying capacity, the population can begin to increase. Factors that can limit population in a habitat are physical factors (temperature, nutrient availability, amounts of light, amount of dissolved oxygen, or pH) and biotic factors (parasites, predators, competitors).
(4 points maximum—2 for definition and 1 for each correct example)

   (b) One example of how nature limits consumption is competition that occurs between two populations for the same habitat. For example, if two different species of animals prey on the same species in the same habitat, they are said to be in direct competition. Sooner or later, the population of prey would be small enough that one predator in the population would not have enough resources. This might cause them to become extinct, to leave that area, or to switch to another food source, thus ending the competition. Some examples of competition are two raptor birds that compete for mice or fish; hunting cats like cheetahs and lions, which compete for grazing animals; or two species of birds that compete for insects.
(4 points maximum—2 points for correctly explaining how consumption can be controlled and 1 point each for 2 correct examples)

   (c) Human activities that violate the limits of population growth can include examples of how we harvest natural resources to help grow food; this affects the habitat of certain plant and animal species. Examples of harvesting more natural resources might include: irrigation to increase water availability; fertilizers to overcome a lack of certain minerals in the soil; or turning the natural biome into farmland to raise more food crops. We eliminate competitors for our food supply and we use medicines to kill parasites. There are numerous possible correct answers to this question.
(2 points maximum—1 point for each correct explanation of how humans remove competition and exploit resources)

CHAPTER 7 DRILL

Multiple-Choice Answers

1. **A** Smelting is a process that separates a desired ore from other materials in mined ore. It is usually accomplished by heating the ore and ladling off the desired molten element.

2. **C** (A), cost-benefit analysis, is the comparison of the benefit of an action relative to the costs of that action. (B), external costs, are the costs that occur after someone purchases something. For example, after you buy a car, the cost of gasoline is an external cost. (D), marginal benefits, are the tradeoff between how much we gain by buying forest land (for example) or using the money to do some other beneficial activity. (E), externalities, are accidental side effects of something we do that affects other people. So, if you drive a car the pollutants you produce can cause acid rain that kills trees in a forest. That is a negative externality. (C), marginal costs, are the costs of each step in a process. Thus, (C) is the correct answer.

3. **D** Clear-cutting is the removal of all trees from an area at the same time. This is typically done in areas that support fast growing trees, like pines.

4. **A** Moderate irrigation with groundwater over a long period of time can cause serious problems, including a significant buildup of salts on the soil's surface, which makes the land unusable for crops; this condition is known as salinization.

5. **D** One of the major motivations for cutting down rainforests is to increase the amount of grazing land for cattle and other farm animals. All of the other options are problems associated with deforestation, but (D) benefits humans.

6. **A** Greenbelts are used in urban planning in order to increase green space and control the growth of cities. They are open or forested areas built at the outer edge of a city. Because no growth is permitted in them, they can increase the quality of life for people living near them. Sometimes satellite towns are built outside the greenbelt and connected to the city by highways and mass transportation methods.

7. **C** The Bureau of Land Management is responsible for the management of federal rangeland.

8. **C** The amount of copper ore in the earth is limited. It is considered a nonrenewable resource. (A), (B), (D), and (E) are considered to be renewable because a renewable resource is one that will be available as long as humans use it in a sustainable manner.

9. **B** In the Tragedy of the Commons, a common resource is used by many people and then becomes depleted as these people do not regulate their consumption of the resource. Some sources say that 75 percent of the world's 200 commercially usable fish are either overfished or are being fished at their maximum sustainable yield.

10. **A** Traditional industrialized agriculture consumes large amounts of energy and other resources. When all aspects are considered (growing crops, processing, and transportation), 17 percent of the U.S.'s total commercial energy use goes into food production. Typically, for every unit of food energy eaten, it takes ten units of fossil fuel energy to prepare and deliver the food.

11. **E**   CITES is the international law that regulates the international trade in endangered species of both plants and animals. This is the correct answer. (A) deals with species only in the United States. (B) is another United States law that deals with marine mammals. (C) is a law that set how the United States will deal with environmental issues. (D), the Resource Conservation and Recovery Act (RCRA), deals with problems created by underground petroleum tanks.

12. **B**   Pesticide resistance is the only answer choice that represents a problem that results from the use of pesticides. Because pesticides have been used in large amounts, some species of pests have evolved traits that allow them to resist the action of pesticides.

13. **C**   The uses of National Parks are restricted to camping, fishing, and boating. Motor vehicles are permitted, but only in designated areas. (D) and (E) describe the permitted uses of National Wildlife Refuge land, and (A) and (B) both describe acceptable uses of National Forest lands.

14. **B**   Sulfuric acid forms as water seeps through mines and carries off sulfur containing compounds. The chemical conversion of sulfur-bearing minerals occurs through a combination of biological (bacterial) and inorganic chemical reactions.

15. **D**   In 1995, most of the world's nations formed the WTO to establish ground rules for international commerce.

16. **D**   A bottom net is pulled along by one or more boats.

17. **C**   Fish pens or oysters raised on wooden racks are a good example of aquaculture.

18. **A**   Drift nets are not towed or dragged. They float freely for up to six months before they're reeled in.

19. **B**   The hooked line can be more than a mile long and is generally dragged behind a boat.

Free-Response Answer

1. Use the checklists below to determine if your responses were correct. We will use checklists like these when there are many different ways that you could have answered the question. However, remember that your answers should be in paragraph format!

(a) Aspects include

Positive aspects	Negative aspects
Low cost to implement this way of watering	Lots of water lost to evaporation
Little technology and training necessary	Not all areas well-adapted to this technique
Low cost to maintain this way of watering	Delivers more water than plants need
	Water distribution is uneven
	Promotes weed growth along with crops

(2 points maximum—1 point for a correct positive aspect and 1 point for a negative aspect)

(b) Alternatives include

- Lining canals (Positive: reduces water lost to infiltration; Negative: expensive to do and uses resources)

- Leveling fields (Positive: water flows where needed; Negative: very expensive to level fields)

- Irrigate at night (Positive: avoids evaporation; Negative: requires careful planning and training)

- Irrigate only when necessary (Positive: less waste; Negative: difficult to time)

- Use drip irrigation (Positive: water drops right to roots; Negative: uses resources of plastic)

- Center pivot: (Positive: low waste as water is sprayed directly on to plants; Negative: equipment is very costly and runs on fossil fuels)
  (3 points maximum—1 point for identifying process, 1 point for correct positive impact, and 1 point for correct negative impact)

(c) Possible negative impacts are

- saltwater intrusion: as the aquifer diminishes, nearby ocean water can migrate underground

- diminished water for domestic or industrial use

- subsidence: as water is withdrawn, the soil settles and sinkholes can develop; these can damage buildings and destroy ecosystems
  (2 points maximum—1 point for naming impact and 1 point for explanation)

(d) Some positive and negative effects are

Positive effects	Negative effects
Production of low-cost electricity	Costly to build
Reservoirs used for many recreational activities	Negative impact on local ecology
Provides flood control	Prevents silt recharging of floodplain
Irrigation water can be controlled	Decrease in fish migration and spawning
Durable	Great danger if breached

(3 points maximum out of 4 possible—1 point for each positive and negative impact)

CHAPTER 8 DRILL

Multiple-Choice Answers

1. **D** Net energy yield is a comparison between cost of extraction, processing, and transportation and the amount of useful energy derived from the fuel. For example, the net energy ration of natural gas used to heat homes is 4.9 and the net energy yield for electrical heating is 0.3. When the two are compared, the net yield from gas is much higher because the costs of getting the gas to your home is very small compared to the costs of running a nuclear power plant.

2. **D** This question asks you to calculate the amount of light produced by a bulb. You know the amount of energy going into the bulb (1.00 joules) and its efficiency (3 percent). So, 3 percent of 1.00 is 0.03. (D) is the correct answer because the useful energy produced is light, not heat.

3. **C** Both methane and ethanol are created as bacteria breaks down biomass. An example of this is seen when farmers produce methane from decomposing manure to heat their barns. The manure is placed in underground pits, where bacteria break it down and release methane.

4. **E** The most common hybrid engines are gasoline-electric hybrids. The two engines work together to provide acceleration and power. When the car is driving slowly, less than 56 kph (35 mph), the electric engine powers the car.

5. **D** The only answer choice that describes a waste of energy is leaving room lights on. Most incandescent light bulbs have an energy efficiency rating of 3–5 percent. So, for every 100 units of energy, we only get 5 units of useful light.

6. **A** The correct formula for this problem is 4,500 tons × (2,000lb/1 ton) × (5,000 BTU/lb) = 4.5 × 1010. It is much easier to use scientific notation here, so (4.5 × 103) × (2 × 103) × (5 × 103) = 45 × 109 or 4.5 × 1010.

7. **C** CO_2 emissions are high in any process that releases energy by burning material. Therefore, burning coal, oil, diesel, and wood all generate CO_2. The processing and reprocessing of uranium into fuel rods generates a moderate amount of CO_2. The only noncombustion generation occurs by wind turbines.

8. **E** WATTS × TIME = kWh. 1,000 watts = 1 kWh, so $\dfrac{500}{1,000} = 0.5$ kWh

9. **B** Photovoltaic cells are constructed from silicon and boron. When sunlight strikes the cells, electrons are energized. These electrons can then flow freely, producing an electric current. By placing wires in the correct positions, this current can be used to power devices. Because the cells are expensive to make, the cost per kilowatt hour is high, but they are useful for applications in which there is no other source of electricity.

10. **C** Radioactive half-life is the amount of time it takes for half of a radioactive sample to disappear. In this question, there are 3 half-lives: the first is 2 curie → 1 curie, the second is 1 curie → 0.5 curie, and the third is 0.5 curie → 0.25 curie. According to the statement, each half-life lasts 20 years. So, 3 × 20 = 60 years.

11. **B** Appliances like microwaves do not have an off switch. They continuously draw power to remain in an "instant on" mode.

12. **D** Coal, oil, and natural gas are burned to generate heat and a nuclear reactor generates heat by radioacive decay. The heat turns water to steam and then steam spins a turbine that in turn spins a generator.

Free-Response Answer

1. **(a)** The core is the place in a nuclear reactor where nuclear reactions occur. It is made of very high strength steel and other high-performance materials. Inside the core are the coolant, fuel rods, and moderators that control the reaction. The fuel rods are made of enriched uranium U-235. In the reactor, a chain reaction generates the heat, which is converted into steam to drive the electricity-producing turbines. The coolant prevents the melting of the uranium, or core, by removing the heat generated by the chain reaction. Some of the heat is used to turn water to steam that spins turbines. In most reactors, the heat produced in the core does not come in contact with the water that is vaporized; instead a device called a heat exchanger takes heat from the core (carried by water or molten sodium) and transfers it to water, which vaporizes.
    (2 points maximum—1 point for each correct description)

    (b) Currently all highly radioactive waste is stored on the grounds of the power plant in deep pools of water. Some materials are stored in steel drums that are housed inside storage containers with walls made of lead and concrete. The United States is currently studying the possibility of opening a deep underground storage at Yucca Mountain, Nevada. Low level radioactive materials are also stored in radioactive landfills.
    (4 points maximum—2 points for a correct name and description of a storage technique)

    (c) Because nuclear power plants do not use the combustion of fossil fuels to generate the heat to produce steam, there are no by-products of combustion. There is no particulate produced, no production of oxides of nitrogen or sulfur, and no release of heavy metals. Some CO_2 is generated, but this occurs primarily in the processing of the fuel rods.
    (2 points maximum—1 point for describing the lowered emissions and 1 point for a correct example)

    (d) Thermal pollution is generated from the turbines, which are powered by steam from the heat exchanger. These hot turbines must be cooled by water piped in from nearby oceans or rivers. This problem is mitigated by allowing the turbine coolant water to flow up a series of pipes built inside cooling towers.

As cool air enters the bottom of the tower, it rises by convection and removes the heat from the water circulating in the pipes. The cooler water is then released.

(2 points maximum—1 point for describing how thermal pollution occurs and 1 point for a method to reduce it.)

CHAPTER 9 DRILL

Multiple-Choice Answers

1. **A** (B) to (E) all contribute directly to the pollution of groundwater. (A) can contribute indirectly if the exhaust is converted into a secondary pollutant (such as acid precipitation).

2. **B** Gray-air smog is a result of the burning of large volumes of coal to generate electricity and provide heat to homes. China uses a great deal of coal to meet its energy needs. The other cities listed have reasonably effective controls on their generation stations, and people in those countries do not use much coal to heat their homes.

3. **D** Sanitary landfills are designed to completely isolate garbage. These landfills contain clay and plastic liners that hold back and collect water as it passes through. The trash is compacted into the smallest area possible and covered with soil and more plastic and clay. As the trash decomposes, the methane generated is collected to be used for other purposes.

4. **E** This process is called eutrophication, and it occurs when organic material (or nutrients) is added to water. The waste supplies food to the bacteria, which thrive using the oxygen for metabolism. This lowers the level of oxygen in the water, which deprives fish and other aquatic animals.

5. **B** The presence of anaerobic organisms, such as bacteria and fungi, indicate that there is little oxygen in a body of water. The presence of trout, perch, and insect larvae would mean that the water was not very polluted. The absence of these indicator species would indicate that the water is polluted.

6. **C** Each person in the United States produces 770 kgs (1,700 lbs) of waste each year. 54 percent of this waste is put in landfills, 30 percent is recycled, and most of the rest is burned (16 percent).

7. **D** The concentration of CO_2 in the atmosphere is increasing more rapidly than that of any other gas. Carbon dioxide is produced during the combustion of fossil fuels and levels of CO_2 are expected to increase rapidly in the next 20 years. Water vapor also contributes to the Greenhouse Effect, but its levels have remained consistent for hundreds of years.

8. **B** In the correct order, first, solids are removed by a series of screens. Next, the water is passed into tanks that contain bacteria. They remove 97 percent of the organic waste. Finally, chlorine is added to kill bacteria and denature viruses.

9. **D** PANs (Peroxyacyl nitrates) are secondary pollutants. They are produced from the reaction of hydrocarbons, oxygen, and nitrogen dioxide. All the remaining options are primary pollutants, which are produced from burning fossil fuels.

10. **B** The repository was being built in Yucca Mountain, Nevada. President Obama has delayed the storage of any radioactive material there.

11. **A** When nitrogen oxides react with water vapor in the atmosphere, the result is nitric acid HNO_3.

12. **C** Urban heat islands are created because of the presence of buildings, highways, factories, and automobiles, and the use of lights warm the surrounding air. This can cause the formation of clouds and can trap pollution near the earth and prevent it from being diluted.

13. **D** Chemicals can be added to bring the pH of hazardous wastes to the neutral seven, which can render some wastes less harmful.

14. **C** Peroxyacyl nitrates are an example of a pollutant that's generated by combustion of wastes. Additionally, hydrocarbons and chlorine can combine during burning to create PANs.

15. **A** Deep wells are created to isolate wastes deep underground. They are cheap and easy to do, but this type of disposal involves the risk of allowing the pollutant to spread into underground aquifers.

16. **B** By reducing the amount of pollutants we produce, we do not have to pay to clean them up or dispose of them. Reduction can occur by changing manufacturing processes or lowering the demand for products that cause waste.

17. **E** In bioremediation, bacteria absorb the material and convert it into other compounds. Bacteria are useful in processing pesticides and diesel fuel.

Free-Response Answer

Use the following lists to check your answers, but remember that you must write all of your answers in essay form!

1. **(a)** Primary pollutants include

    Oxides of carbon—CO and CO_2

    Oxides of nitrogen—NO and NO_2

    Oxides of sulfur—SO_2

    Unburned hydrocarbons

    All of these primary pollutants are produced by the burning of fossil fuels.
    (3 points maximum—1 point for each correct pollutant and 1 point for correct origin)

    (b) Secondary pollutants include

    Sulfur trioxide—SO_3

    Nitric acid—HNO_3

    Sulfuric acid—H_2SO_4

Hydrogen peroxide—H_2O_2

Ozone—O_3

Peroxyacyl nitrates or PANs—$R\text{-}C(O)OONO_2$

Aldehydes—$R\text{-}COH$

All of these secondary pollutants are formed in the atmosphere. When primary pollutants combine with water vapor and sunlight energy, the reactions that take place produce the above products.
(3 points maximum—1 point for each correct pollutant and 1 point for a correct explanation of how it's produced)

(c) Photochemical smog is more likely to be found in industrialized nations because of their extensive use of fossil fuels (mostly coal and oil). Gray-air smog is found in areas where coal is the dominant fossil fuel.
(2 points maximum—1 point for photochemical smog explanation and 1 point for gray-air smog explanation)

(d) Ozone causes breathing problems, eye irritation, coughing, and reduced immune response. It also aggravates chronic diseases. CO can reduce the oxygen-carrying capacity of blood and cause dizziness and nausea. It can also retard fetal development. NO_2, SO_2, and PANs cause breathing problems, aggravate existing breathing problems, and increase susceptibility to disease.
(2 points maximum—1 point for irritant and 1 point for human effect)

CHAPTER 10 DRILL

Multiple-Choice Answers

1. **A** The Montreal Protocol was signed in 1987 by 36 nations concerned with the depletion of the ozone layer by chlorofluorocarbons (CFCs). The pact stated that CFCs had to be reduced 35 percent between 1989 and 2000. There was a further refinement of the agreement in Copenhagen, Denmark in 1992.

2. **B** HMTA was passed in 1975 with the intention of defining what a hazardous material was and setting rules for Hazmat shipment.

3. **D** The Marine Mammal Protection Act, established in 1972, is legislation that protects marine mammals in the world's oceans.

4. **D** Large federal projects that might have a large impact on the environment must produce an environmental impact statement. This is mandated by the 1969 National Environmental Policy Act.

5. **A** In 1962, Rachel Carson published the book *Silent Spring*. In this book she describes the problems associated with the overuse of pesticides—mostly DDT. She explains how pesticides were affecting bird populations in the United States. The book, along with her advocacy, changed the way people thought about the impact of pesticides.

6. **E** The CAA enabled 110 of the most polluting power plants in 21 states to buy and sell pollution rights. This regulated how much pollution (SO_2) each plant could produce. If the plant produces an amount

under the level, it receives a "credit" which can be kept or sold to another plant that is producing pollution over the limit of permitted emissions.

7. **E** Environmental economics is the study of the impact of the goods and services economy, particularly market systems of allocation on environmental quality and ecological integrity. It also takes into consideration the economic basis for pollution problems and policy alternatives. Discussions of environmental issues always include monetary and economic considerations. For example, what are the relative costs and benefits of proposed policies?

8. **B** The Endangered Species Act was passed by Congress in 1973 to provide protection for species that are threatened with extinction. This act authorizes Federal Agencies to undertake conservation programs to protect species listed as endangered or threatened, and to purchase land to protect habitats.

9. **D** Holistic viewpoints center on the belief that we are one of many species on the planet, and that we interact with all species. They also believe that if we harm the planet, we harm ourselves. The other options are all "planet management" viewpoints.

10. **C** Sustainability depends on the long-term utilization of resources; this is described in (C). The other options could cause resources to be used up more rapidly.

11. **E** A cap-and-trade policy limits carbon dioxide in the atmosphere by implementing caps and offering incentives for reducing emissions.

12. **A** WWF is an organization that is not associated with any branch of government.

Free-Response Answer

1. **(a)** The CAA sets two standards: National Ambient Air Quality Standards for six outdoor pollutants and the national emissions standards for over 100 toxic pollutants. The NAAQS sets limits on pollutants that people or the environment may be exposed to over a certain period of time. Carbon monoxide, ozone, sulfur dioxide, and nitrogen dioxide are compounds of concern to people near the plant. The second set of standards describes specific limits of hazardous air pollutants. Examples include lead, mercury, and radionuclides.
(2 points maximum—1 to describe the goal of each standard [NAAQS & hazardous air pollutant] and 1 for a correct example for each)

(b) Positive results include more employment, reliable electricity, lower cost electricity, less dependence on foreign energy sources, greater tax reviews, and a positive impact on the local economy.

Negative impacts include more air pollution, more water pollution (thermal is an example), the issue of ash disposal, the need to build infrastructure to support the plant (railroads for coal, roads to and from the plant), and the loss of the natural beauty of the area.
(4 points maximum—1 point for each positive and each negative impact)

(c) ESA could be used if a threatened or endangered species is found in or near the area where the plant is to be built. The act allows for the conservation of ecosystems that support the endangered or threatened species. It is possible that the plant would not be built in order to conserve the ecosystem in which the species lives.

(2 points maximum—1 point for defining the goal of the ESA and 1 point for stating that the plant could not be built)

(d) PPA is designed to promote source reduction (stopping pollution from being produced). Sulfur reductions could occur by burning low-sulfur coal or by using fluidized-bed coal combustion. The sulfur and nitrogen oxides can be removed by mixing coal and limestone and burning it in a large volume of air. Another process is to wash the coal with water and remove pyritic sulfur particles.

(2 points maximum—1 for correctly stating the PPA goal and 1 for a correct example of how to reduce pollutants before they are made. Note: No credit would be given for examples like scrubbers or catalysts because they clean up the pollutants after they are produced. PPA focuses on ways to lower the amounts produced, not how to clean them up after production.)

Part V
Practice Tests

Chapter 14
Practice Test 1

AP® Environmental Science Exam

DO NOT OPEN THIS BOOKLET UNTIL YOU ARE TOLD TO DO SO.

At a Glance

Total Time
1 hour and 30 minutes
Number of Questions
100
Percent of Total Grade
60%
Writing Instrument
Pencil required

Instructions

Section I of this examination contains 100 multiple-choice questions. Fill in only the ovals for numbers 1 through 100 on your answer sheet.

Indicate all of your answers to the multiple-choice questions on the answer sheet. No credit will be given for anything written in this exam booklet, but you may use the booklet for notes or scratch work. After you have decided which of the suggested answers is best, completely fill in the corresponding oval on the answer sheet. Give only one answer to each question. If you change an answer, be sure that the previous mark is erased completely. Here is a sample question and answer.

Sample Question   Sample Answer

Chicago is a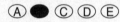

(A) state
(B) city
(C) country
(D) continent
(E) village

Use your time effectively, working as quickly as you can without losing accuracy. Do not spend too much time on any one question. Go on to other questions and come back to the ones you have not answered if you have time. It is not expected that everyone will know the answers to all the multiple-choice questions.

About Guessing

Many candidates wonder whether or not to guess the answers to questions about which they are not certain. Multiple choice scores are based on the number of questions answered correctly. Points are not deducted for incorrect answers, and no points are awarded for unanswered questions. Because points are not deducted for incorrect answers, you are encouraged to answer all multiple-choice questions. On any questions you do not know the answer to, you should eliminate as many choices as you can, and then select the best answer among the remaining choices.

GO ON TO THE NEXT PAGE.

ENVIRONMENTAL SCIENCE
Section I
Time—One hour and 30 minutes
Part A

Directions: Each set of lettered choices below refers to the numbered questions or statements immediately following it. Select the one lettered choice that best answers each question or best fits each statement and then fill in the corresponding oval on the answer sheet. A choice may be used once, more than once, or not at all in each set.

Questions 1-5 refer to the structure of the atmosphere.

    (A) Troposphere
    (B) Stratosphere
    (C) Thermosphere
    (D) Mesosphere
    (E) Stratopause

1. The layer that contains the earth's daily weather

2. Extends from 50–85 km above Earth

3. The earth's ozone layer exists in this layer of the atmosphere

4. The highest layer of the atmosphere (above 80 km)

5. The layer of the atmosphere that's heated by infrared radiation from the earth

Questions 6-10 refer to the following soil layers.

    (A) A horizon
    (B) B horizon
    (C) E horizon
    (D) O horizon
    (E) Bedrock

6. Silt and sand are concentrated here

7. Litter layer, mostly undecayed materials

8. The deep, underlying non-soil materials

9. The layer where minerals that were leached out of layers above accumulate

10. A mixture of soil, loam, and detritus; the topsoil

GO ON TO THE NEXT PAGE.

Questions 11-15 refer to the following five age-structure pyramids.

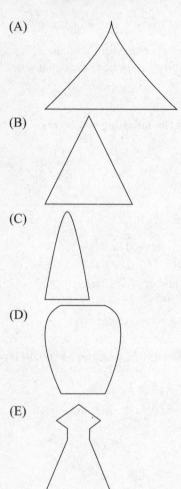

(A)

(B)

(C)

(D)

(E)

11. A country that is growing slowly

12. A country at zero population growth

13. A country that is losing many of its young adults to diseases like AIDS

14. A rapidly growing population

15. A country showing a population decline

Questions 16-20 refer to the following risks to human health.

(A) Radon
(B) Asbestosis
(C) Malaria
(D) Earthquake
(E) AIDS

16. The virus that causes this disease is transmitted through bodily fluids

17. Can cause massive destruction

18. Caused by microscopic fibers of a mineral

19. Radiation that causes lung cancer

20. Caused by a protozoan carried by mosquitoes

GO ON TO THE NEXT PAGE.

Questions 21-25 refer to the following soil types.

    (A) Desert soil
    (B) Grassland soil
    (C) Tropical rain forest soil
    (D) Pine forest soil
    (E) Deciduous forest soil

21. Soil that has a substantial organic layer; fire helps to break down plant material in this layer

22. Soil composed of litter and humus; this soil is acidic due to the accumulation of needles

23. Soil is rocky, very dry, and contains almost no organic matter

24. Soil is acidic and contains very little organic matter despite large plant populations

25. Soil is rich in humus and partially decayed leaves

Questions 26-30 refer to the following atmospheric pollutants.

    (A) Carbon monoxide
    (B) Nitrogen dioxide
    (C) Sulfur dioxide
    (D) Photochemical oxidant
    (E) Suspended particulate matter

26. Is involved in the formation of nitric acid

27. Dust or soot

28. Ozone

29. Is involved in the formation of sulfuric acid

30. Health effects include reduced blood oxygen levels

GO ON TO THE NEXT PAGE.

Part B

Directions: Each of the questions or incomplete statements below is followed by five suggested answers or completions. Select the one that is best in each case and then fill in the corresponding oval on the answer sheet.

31. The goal of the second stage of a waste water treatment plant is to

    (A) remove the large solid material
    (B) aerate the water
    (C) make muddy water clear
    (D) remove chemicals such as DDT or PCBs
    (E) lower the amount of organic material in the water

32. Which of the following organisms is the first to be adversely affected by thermal pollution in a stream?

    (A) Trees along the bank
    (B) Insect larvae in the water
    (C) Large fish migrating up stream
    (D) Birds drinking the water
    (E) Bacteria in the water

GO ON TO THE NEXT PAGE.

Questions 33-36 refer to the following passage and graph.
A scientist placed 100 fish eggs into each of seven solutions with different pH values. After 96 hours the number of survivors was counted and converted into a percent. The percent surviving is given in the graph below.

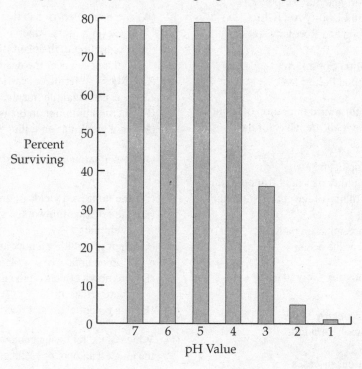

33. Which of the values below best represents the LD_{50} in this experiment?

   (A) 6.0
   (B) 4.0
   (C) 4.5
   (D) 3.0
   (E) 2.5

34. At what pH value do the fewest fish hatch?

   (A) 7.0
   (B) 6.0
   (C) 3.5
   (D) 2.0
   (E) 1.0

35. Which of the following best describes the goal of the above experiment?

   (A) To test the hypothesis that the bigger the fish, the smaller the pH tolerance range.
   (B) To observe how many fish would hatch at different pH values.
   (C) To find out how many fish live in streams with different pH values.
   (D) To understand how acid rain affects life in streams.
   (E) To see what chemical is best at changing the pH of water.

36. The pH value is a measure of the

   (A) amount of heavy metals in the water
   (B) BOD of the water
   (C) concentration of oxygen in the water
   (D) concentration of hydrogen ions in the water
   (E) depth the scientist can see under the water

GO ON TO THE NEXT PAGE.

37. Which of the following laws created the Superfund program?

    (A) Comprehensive Environmental Response, Compensation, and Liability Act (CERCLA)
    (B) Resource Conservation and Recovery Act
    (C) Clean Air Act
    (D) Federal Water Pollution Control Act
    (E) National Environmental Policy Act

38. Later fall frosts and the northward migration of some tree and plant species may indicate which of the following global changes?

    (A) Increased global temperatures
    (B) The effects of more ultraviolet light from the sun
    (C) A reduction in the volume of ice at the North and South Poles
    (D) Changes in global precipitation patterns
    (E) Flooding of areas near the ocean

39. High infant mortality rates are likely to occur in countries that have

    (A) a strong and stable economy
    (B) high levels of education for adults
    (C) a stable food supply
    (D) high levels of infectious diseases
    (E) safe drinking water

40. Composting is a process that produces

    (A) useful plastic products
    (B) a nutrient-rich soil conditioner
    (C) manure
    (D) lower-grade paper products
    (E) materials used in construction

41. All of the following statements are true EXCEPT

    (A) Energy can be converted from one form to another.
    (B) Energy input always equals energy output.
    (C) Energy and matter can generally be converted into each other.
    (D) The laws of thermodynamics can be applied to living systems.
    (E) At each step of an energy transformation, some energy is lost to heat.

42. Oxygen-depleted zones of the oceans, such as the one at the mouth of the Mississippi River, are most likely caused by

    (A) large numbers of fish that are using up all the oxygen in the water
    (B) a reduction in the plant life in rivers that empty into the ocean near the dead zone
    (C) excessive fertilizers carried into the ocean, which cause algal blooms that lower the oxygen levels
    (D) thermal pollution in the ocean
    (E) acid precipitation falling on the ocean

43. One potential benefit to using genetically modified foods is

    (A) the improved yields of crops
    (B) the release of unwanted genes to other plants or animals
    (C) their growth in monoculture will reduce biodiversity in an area
    (D) unknown effects on the ecosystem into which they are released
    (E) the potential greater need for fertilizers

44. Which of the following compounds would probably supply the greatest amount of useful energy to humans?

    (A) The exhaust from a car
    (B) Unrefined aluminum ore
    (C) A glass bottle
    (D) Heat used to warm a home
    (E) A liter of gasoline

45. Which of the following choices gives the geologic eras in the correct sequence, from the oldest to the most recent?

    (A) Cenozoic—Mesozoic—Paleozoic—Precambrian
    (B) Precambrian—Paleozoic—Mesozoic—Cenozoic
    (C) Paleozoic—Precambrian—Cenozoic—Mesozoic
    (D) Paleozoic—Cenozoic—Precambrian—Mesozoic
    (E) Mesozoic—Paleozoic—Precambrian—Cenozoic

46. Which of the following figures most accurately gives the percent of the world's solid waste produced by the United States?

    (A) 50 percent
    (B) 40 percent
    (C) 33 percent
    (D) 10 percent
    (E) 5 percent

GO ON TO THE NEXT PAGE.

47. Which of the following correctly describes conservation easement?

    (A) It is a process that conserves soil from erosion.
    (B) This is a binding agreement that preserves land from further development in exchange for tax write-offs.
    (C) This agreement allows a developer to add new land to a housing project with little input from neighbors.
    (D) This practice prevents the breakdown of stream banks.
    (E) This is a method of building a landfill to minimize runoff.

48. The highest priority of the Clean Water Act is to provide

    (A) funds to increase recycling participation
    (B) guidance in toxic chemical disposal
    (C) funds to reclaim old strip mines
    (D) policies to lessen the amount of oil spills in the ocean
    (E) policies to attain fishable and swimmable waters in the United States

49. Which of the following best describes changes in the genetic composition of a population over many generations?

    (A) Evolution
    (B) Mutation
    (C) Natural selection
    (D) Emigration
    (E) Biomagnification

50. Women have fewer and healthier children when all of the following are true EXCEPT

    (A) they have little education
    (B) they live where their rights are not suppressed
    (C) they have access to medicine and health care
    (D) the cost of a child's education is high
    (E) they have access to birth control

51. "The maximum number of a species that can be sustained in an ecosystem." This phrase best defines

    (A) the carrying capacity
    (B) an ecotone
    (C) the upward curve of a population graph
    (D) natural selection
    (E) a community

52. An increase in the amount of UV light striking the earth as a result of ozone loss will cause which of the following?

    (A) Global climate change
    (B) Increased skin cancer rates in humans
    (C) Lowering of ocean water levels
    (D) An increase in CO_2 in the atmosphere
    (E) A change in the North Atlantic Current

53. Ozone in the troposphere can result in all of the following EXCEPT

    (A) eye irritation
    (B) lung cancer
    (C) bronchitis
    (D) headache
    (E) emphysema

54. Which of the following describes the amount of energy that plants pass on to herbivores?

    (A) The amount of solar energy in a biome
    (B) The First Law of Thermodynamics
    (C) The Net Primary Productivity (NPP) of an area
    (D) The Second Law of Thermodynamics
    (E) The number of steps in the food web

55. The second law of thermodynamics relates to living organisms because it explains why

    (A) matter is never destroyed but it can change shape
    (B) living cells come from other living things
    (C) plants need sunlight in order to survive
    (D) all living things must have a constant supply of energy in the form of food
    (E) the amount of energy flowing into an ecosystem is the same as the amount flowing out of that system

56. Acid deposition most severely affects amphibian species because amphibians

    (A) do not care for their young
    (B) are not mammals
    (C) need to live in both terrestrial and aquatic habitats
    (D) seldom reproduce
    (E) eat only small insects

GO ON TO THE NEXT PAGE.

57. All of the following are internal costs of an automobile EXCEPT

    (A) car insurance
    (B) fuel
    (C) pollution and health care costs
    (D) raw materials and labor
    (E) new tires

58. Scrubbers are devices installed in smoke stacks to

    (A) reduce the amount of materials such as SO_2 in the smoke they discharge
    (B) clean out the stack so smoke can move rapidly upwards
    (C) reduce the amount of sulfur in coal before it is burned
    (D) clean out the boilers for more efficient operation
    (E) reduce the amount of toxic ash produced

59. After ore is mined, the unusable part that remains is placed in piles called

    (A) overburden
    (B) seam waste
    (C) leachate
    (D) tailings
    (E) reclamation

60. All of the following are economic goods EXCEPT

    (A) a swing set
    (B) computer repair service
    (C) food
    (D) a walk in the woods
    (E) a ticket to a game

61. Which fuel contains the greatest amount of sulfur?

    (A) Wood
    (B) Natural gas
    (C) Oil
    (D) Nuclear reactor fuel rods
    (E) Coal

62. Which of the following items includes the others?

    (A) Renewable resources
    (B) Natural resources
    (C) Economic resources
    (D) Manufactured capital
    (E) Labor

63. Biological reserves are areas that allow countries to

    (A) concentrate agricultural production in one area
    (B) set aside critical habitats to ensure the survival of species
    (C) control the flow of rivers and storm waters
    (D) provide grazing land in order to ensure economic growth
    (E) obtain needed minerals from underground mines

64. Which of the following countries has the largest population?

    (A) Japan
    (B) United States
    (C) Brazil
    (D) China
    (E) Russia

65. Which of the following phases of the hydrologic cycle requires the input of solar energy?

    (A) Percolation
    (B) Bioremediation
    (C) Precipitation
    (D) Condensation
    (E) Evaporation

66. Full cost pricing of a refrigerator would include

    (A) adding the cost of employee salaries to the total cost
    (B) the refrigerator's total impact on the environment
    (C) the cost of transporting the refrigerator to the retail store
    (D) the value of the refrigerator if it was donated to a nonprofit group
    (E) changing the color of the refrigerator at a later date

67. A certain chemical has a concentration of 10 ppm in water. Which statement most accurately describes its concentration?

    (A) There are 10 molecules of the chemical in one million molecules of water.
    (B) There are 10 million molecules of the chemical in the sample.
    (C) There are 10 million molecules of the chemical in a 1-liter beaker.
    (D) There are 10 molecules of water in one million molecules of the chemical.
    (E) There are 10 molecules of the chemical in 10 million molecules of water.

GO ON TO THE NEXT PAGE.

68. During a society's postindustrial state, the population will exhibit

    (A) rapid growth with a low birth rate and high death rate
    (B) slow rate of growth with slowing birth rate and a low death rate
    (C) a rapid growth rate with high birth rate and low death rate
    (D) a zero growth rate with a low birth rate and a low death rate
    (E) a slow rate of growth with a high birth rate and a high death rate

69. Which of the following is NOT a disadvantage of old-style landfills?

    (A) They generate gases that can be recovered and used as fuel.
    (B) Bad odors come from these landfills.
    (C) Toxic wastes leach into ground water.
    (D) Subsidence of the land after the landfill is filled.
    (E) They create an eyesore in the neighborhood.

70. The international treaty concerning endangered species (CITES) has tried to protect endangered species by which of the following steps?

    (A) Making more countries keep these species in zoos
    (B) Paying the debts of member countries in order to relieve the pressure to sell endangered species
    (C) Developing a list of endangered species and prohibiting trade in those species
    (D) Providing member countries with a police force to uphold the CITES treaty
    (E) Restoring endangered habitats

71. Country A had a birth rate of 12 per 1,000 in 2000 and a death rate of 9 per 1,000 in the same year. Which of the following is the correct rate of growth for the year 2000?

    (A) 36.0 percent
    (B) 27.0 percent
    (C) 4.0 percent
    (D) 3.0 percent
    (E) 0.3 percent

72. Salt intrusion into freshwater aquifers; beach erosion; and the disruption of costal fisheries are all possible results of which of the following?

    (A) Rising ocean levels as a result of global warming
    (B) More solar ultraviolet radiation on the earth
    (C) More chlorofluorocarbons in the atmosphere
    (D) Reduced rates of photosynthesis
    (E) Use of the oceans as waste disposal area

73. The chemical actions that produce compost would best be described as

    (A) photosynthesis
    (B) augmentation
    (C) respiration
    (D) decomposition
    (E) nitrification

74. Which of the sources below would produce non-point source pollution?

    (A) The smoke stack of a factory
    (B) A volcano
    (C) A pipe leading into a river from a sewage treatment plant
    (D) A car's exhaust pipe
    (E) A large area of farmland near a river

75. A nation's gross domestic product describes

    (A) the amount of public transportation
    (B) the ability to provide health care
    (C) the amount of goods it imports
    (D) the amount of its economic output
    (E) the quality of its environment

76. Which of the following mining operations requires people and machinery to operate underground?

    (A) Mountain top removal
    (B) Contour stripping
    (C) Dredging
    (D) Area stripping
    (E) Shaft sinking

77. A country's total fertility rate (TFR) best expresses which of the following?

    (A) The life expectancy of women in the country
    (B) The average number of babies born to women between the ages of 14 and 45
    (C) The total economic value of all foreign and domestic services
    (D) The number of babies under one year of age who die per 1,000
    (E) The total use of contraceptives in the country

GO ON TO THE NEXT PAGE.

78. The wastes stored in Love Canal contaminated the surrounding area by all of the following methods EXCEPT

    (A) leaching into the ground water
    (B) fumes from burning the wastes
    (C) flowing in the sewers
    (D) runoff into a nearby stream
    (E) spilled drums of waste

79. The distinct building blocks of matter are called

    (A) mixtures
    (B) isotopes
    (C) atoms
    (D) electrons
    (E) compounds

80. In sea water, carbon is mostly found in the form of

    (A) phosphoric acid
    (B) carbon disulfide
    (C) bicarbonate ions
    (D) methane gas
    (E) glucose

81. Acid rain and snow harm some areas more than other areas because certain areas

    (A) have more bacteria in the soil than others
    (B) have less of an ability to neutralize the acids
    (C) are at a higher elevation than the unaffected areas
    (D) are closer to lakes than the unaffected areas
    (E) have more complex food webs than the unaffected areas

82. The one area that does NOT store a lot of phosphorus is

    (A) rocks
    (B) water
    (C) atmosphere
    (D) living organisms
    (E) guano (bird droppings)

83. The addition of oxygen to the early earth's atmosphere most likely occurred through the process of

    (A) volcanic outgassing
    (B) photosynthesis
    (C) meteorite impact
    (D) respiration by animals
    (E) bubbling geysers

84. Which processes do scientists use to estimate environmental risks to humans?

    I. Animal studies
    II. Epidemiological studies
    III. Statistical Probabilities

    (A) I only
    (B) II only
    (C) I and II only
    (D) I and II only
    (E) I, II, and III

GO ON TO THE NEXT PAGE.

Questions 85-88 refer to the following graph.

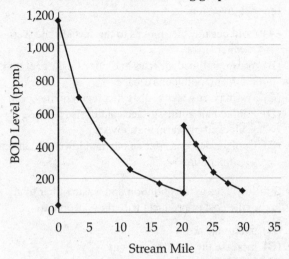

A group of students did a biological oxygen demand study along a 30-mile section of a stream. The data they obtained is given in the graph above.

85. Which of the following best describes the type of pollution at mile 0?

(A) Point source
(B) Thermal inversion
(C) Acid deposition
(D) Secondary pollutant
(E) Deep well

86. The BOD at mile 12 is approximately

(A) 700 ppm
(B) 220 ppm
(C) 200 ppm
(D) 175 ppm
(E) 50 ppm

87. The BOD test is designed to directly measure

(A) how much light can pass to the bottom of the stream
(B) the amount of nitrates in the water
(C) how rapidly the water is moving
(D) the amounts of coliform bacteria
(E) the rate at which oxygen is being consumed by microorganisms

88. Anaerobic bacteria, sludge worms, and fungi are most likely to be found in which part of this stream?

(A) 0 to 5 miles
(B) 10 to 15 miles
(C) 25 to 30 miles
(D) 10 to 20 miles
(E) 15 to 20 miles

GO ON TO THE NEXT PAGE.

89. Riparian zones are important parts of lands because they are

    (A) the area where most cattle feed when they graze
    (B) an area of diverse habitats along the banks of rivers
    (C) important buffers against wind
    (D) areas where varying amounts of light cause different layers of plant growth
    (E) the origins of rivers

90. Which of the following is a disadvantage of fish farming?

    (A) It can produce large volumes of fish for food.
    (B) It can allow for genetic engineering, which leads to bigger yields.
    (C) It is very profitable.
    (D) It can lead to large die-offs due to disease.
    (E) It can reduce the pressure to harvest wild species.

91. Which of the following philosophies would be advocated by someone with the "environmental wisdom" point of view?

    (A) As the planet's dominant species, we are most important.
    (B) All economic growth is good.
    (C) Society can use resources at an uncontrolled pace.
    (D) We will do best when humans manage the planet.
    (E) All species are important and we are not in charge.

92. The form of nitrogen that plants can use directly is

    (A) nitrates
    (B) nitrites
    (C) guano
    (D) N_2 gas
    (E) methane

93. Which of the following best describes the effects of a thermal inversion?

    (A) Cold ocean water moves to the surface and warm water sinks.
    (B) Warm, polluted air rises and mixes with cool upper air, and pollutants escape.
    (C) Warm river water cools when it enters the ocean.
    (D) Polluted air at the surface cannot rise because it is blocked by warm air above it.
    (E) Cool air descends onto a city and lowers nighttime temperatures.

94. Shifting taxes to tax pollution and waste rather than taxing the cost of products will allow people to

    (A) maximize profit
    (B) increase the tax base in a city
    (C) hold industry more accountable for pollution
    (D) shift to a pattern of more sustainable development
    (E) keep the cost of collecting taxes down

95. Which of the following molecules is most damaging to stratospheric ozone?

    (A) H_2O
    (B) CO_2
    (C) Chlorofluorocarbons
    (D) N_2O
    (E) SO_2

GO ON TO THE NEXT PAGE.

96. Which of the following series of numbers demonstrates exponential growth?

    (A) 200, 199, 198, 197, 196…
    (B) 1, 3, 5, 7, 9…
    (C) 2, 4, 8, 16, 32…
    (D) 1, 3, 9, 27, 81…
    (E) 2, 4, 6, 8, 10…

97. Samples of atmospheric gases from past eras can most easily be obtained from which of the following sources?

    (A) Methane gas trapped in oil reserves
    (B) Different types of sedimentary rock
    (C) Gases trapped in polar ice caps
    (D) Tree ring measurements
    (E) Mud samples from eutrophic lakes

98. Acid deposition on soil kills beneficial decomposers; which of the following cycles would be most affected by the loss of decomposers?

    (A) Sulfur cycling
    (B) Phosphorus cycling
    (C) Hydrologic cycling
    (D) Nitrogen cycling
    (E) Temperature cycling

99. Which of the following is a trace element necessary for plant growth?

    (A) Carbon
    (B) Nitrogen
    (C) Phosphorous
    (D) Magnesium
    (E) Potassium

100. Concerns that people of color and poor people are unevenly exposed to environmental pollution are most likely to be addressed by people who believe in the

    (A) earth stewardship view
    (B) planetary manager view
    (C) ecofeminist point of view
    (D) the environmental justice movement
    (E) sustainability point of view

END OF SECTION I

GO ON TO THE NEXT PAGE.

ENVIRONMENTAL SCIENCE
SECTION II
Time—90 minutes
4 Questions

Directions: Answer all four questions, which are weighted equally; the suggested time is about 22 minutes for answering each question. Where calculations are required, clearly show how you arrived at your answer. Where explanation or discussion is required, support your answers with relevant information and/or specific examples.

1. According to the United States Energy Information Administration, the consumption of natural gas by the United States increases at 8 percent per year. It receives its supplies from a variety of international and domestic locations. Natural gas is used in the home, industry, and in the generation of power.

   (a) Calculate the approximate number of years it would take to double the consumption of natural gas in the United States. Show all work.

   (b) Describe one method by which natural gas is recovered and transported.

   (c) Describe two benefits to the environment that would occur if the United States switched from coal to natural gas-fired electric power generation.

   (d) Some people advocate increasing the use of coal versus natural gas for the production of electricity. Explain one argument that the proponents of coal might use to justify their position.

   (e) Discuss one non-hydrocarbon fuel alternative and describe ONE of its drawbacks.

2. The map below shows two cities: City *X* and City *Z*, separated by several kilometers.

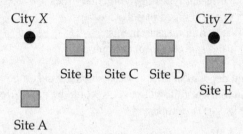

Students from a high school in between the two cities studied soil pH values at the sites labeled A through E on the map. The results of the pH study are given in the following table:

Site	pH value
A	6.2
B	5.6
C	5.0
D	4.5
E	4.3

   (a) Describe one point source for the pollution that caused the change in the soil's pH as shown. Include in the description a fuel that could create the pollution.

   (b) Identify one primary and one secondary pollutant that can cause the change in the soil's pH. Describe the process that causes the change in the pH.

   (c) Describe one possible method to reduce the air pollutants that are causing the pH change.

   (d) Describe one provision of the Clean Air Act of 1990 that could be used to control and reduce the emissions.

GO ON TO THE NEXT PAGE.

3. The following editorial is excerpted from a recent edition of the *Hilltop Express*:

Hilltop Express

New Pests Invade Farm

A new species of corn-infesting insect has recently been discovered in a local farmer's field. Bill Jones stated: "Last week a section of my corn field was covered in small black beetles. They can fly from plant to plant, and they eat large holes in the leaves. I called the county extension agent Sarah Smith and she came out and identified them. I'm going to start spraying tomorrow morning." In a telephone interview with Sarah, she stated that this species was new to the county and has the potential for causing real damage to the corn crop. She stated that the adults do most of the damage to growing leaves.

The grubs live near the base of the plant and feed on bacteria and other organisms living in the soil. She added that the beetle was resistant to the most common pesticide, NOGrub. NOGrub, she commented, had been tried in another county and was not found to be effective. The editors of the Hilltop Express realize the potential dangers to the county's most important cash crop. We urge the county agents to recommend a series of new pesticide treatments to control this new menace to our livelihood.

(a) Describe how the beetle might have become resistant to NOGrub. Assume that NOGrub had been applied to a population of beetles in another county.

(b) Discuss two negative impacts of using chemical pesticides on the surrounding ecosystem.

(c) Describe two other methods of controlling the beetles without resorting to human-made chemical pesticides.

(d) One strategy for combating pests is Integrated Pest Management (IPM).
 (i) Describe IPM.
 (ii) Explain one benefit and one difficulty in using IPM to control this outbreak.

4. Many endangered species live in areas where biodiversity has been degraded by human activities. Species such as the West Virginia spring salamander or the California condor live in areas where the impact of human activities has made these and other organisms very rare.

(a) Discuss two human activities that cause species to become endangered.

(b) Describe two reproductive strategy characteristics that make a species prone to extinction.

(c) Describe one economic and one legislative action that attempt to save endangered species.

STOP
END OF EXAM
_____

Chapter 15
Practice Test 1:
Answers and
Explanations

ANSWER KEY

1.	A	35.	B	69.	A
2.	D	36.	D	70.	C
3.	B	37.	A	71.	E
4.	C	38.	A	72.	A
5.	A	39.	D	73.	D
6.	C	40.	B	74.	E
7.	D	41.	C	75.	D
8.	E	42.	C	76.	E
9.	B	43.	A	77.	B
10.	A	44.	E	78.	B
11.	B	45.	B	79.	C
12.	C	46.	C	80.	C
13.	E	47.	B	81.	B
14.	A	48.	E	82.	C
15.	D	49.	A	83.	B
16.	E	50.	A	84.	E
17.	D	51.	A	85.	A
18.	B	52.	B	86.	B
19.	A	53.	B	87.	E
20.	C	54.	C	88.	A
21.	B	55.	D	89.	B
22.	D	56.	C	90.	D
23.	A	57.	C	91.	E
24.	C	58.	A	92.	A
25.	E	59.	D	93.	D
26.	B	60.	D	94.	D
27.	E	61.	E	95.	C
28.	D	62.	C	96.	C
29.	C	63.	B	97.	C
30.	A	64.	D	98.	D
31.	E	65.	E	99.	D
32.	B	66.	B	100.	D
33.	D	67.	A		
34.	E	68.	D		

Once you have checked your answers, remember to return to page 4 and respond to the Reflect questions.

EXPLANATIONS FOR SECTION I—PART A

1. **A** The troposphere starts at the earth's surface and extends 8–14.5 km (5–9 miles) high; this layer of the atmosphere contains the earth's daily weather.

2. **D** The mesosphere starts just above the stratosphere (50 km) and extends to a distance of approximately 85 km (53 miles) out from Earth.

3. **B** The ozone layer, which absorbs the solar ultraviolet radiation, is in the stratosphere layer.

4. **C** The thermosphere starts just above the mesosphere and extends to 600 km (372 miles) high; it is the layer of the atmosphere that lies farthest from Earth.

5. **A** The troposphere is in contact with the earth. After absorbing light from the sun, the earth reradiates the energy as infrared heat. This heat warms the troposphere and keeps the climate of Earth moderate near its surface.

6. **C** The E horizon, or subsurface layer, is generally bleached (white) in appearance. As water moves down through this horizon, soluble minerals and nutrients in the horizon dissolve and are washed (leached) out of it. This horizon is characterized by the loss of silicate clay, iron, aluminum, humus, or some combination of these, which leave behind concentrated sand and silt particles. If you didn't know what the E horizon is, you could use POE to get this answer anyway. In fact, questions like these are built for POE!

7. **D** The uppermost layer generally is an organic horizon, or O horizon. The O horizon consists of fresh and decaying plant residue from sources like leaves, needles, twigs, moss, lichens, and other organic material accumulations.

8. **E** The lowest horizon, the R horizon, is also known as bedrock. Bedrock can lie within a few inches of the surface, or it may be many feet below the surface. It is composed of large, intact rocks. This material eventually weathers into the gravel that makes up the layers above the bedrock.

9. **B** Below the E horizon is the B horizon, or subsoil. The B horizon is usually lighter colored, denser, and lower in organic matter than the A horizon. It is usually the zone where the materials that were leached from the layers above accumulate.

10. **A** Below the O horizon is the A horizon, also known as topsoil. The A horizon is mainly comprised of mineral material. It is generally darker than the lower horizons because it contains large amounts of decomposed organic matter.

11. **B** In slow-growth populations, the base of the age-structure pyramid is not very broad; this is an indication of few young children (0–14 years old) in the population.

12. **C** In a country with zero population growth, the number of children is at replacement level. Just enough children are born to replace the parents; in the age-structure pyramid for this type of population, the width of the pyramid would be similar along most of its height.

13. **E** In this age-structure pyramid, there is a pinched-in area that indicates that many young adults are dying of disease. In a pyramid that shows a population affected by AIDS, the age group affected is the reproductive age group (15–45 years old).

14. **A** The base of an age-structure pyramid that shows a population with a rising growth rate will be very broad, and this is shown in pyramid A. The broad base indicated the relatively large number of children in the population. As these children age, they will become the parents of the next generation, and because at that point there will be a large number of parents, this population will continue to grow.

15. **D** The base of this age-structure pyramid is very small; this indicates that there are very few children being born in this population—not enough to replace their parents. This will lead to an overall decline in the size of the population.

16. **E** A virus called the human immunodeficiency virus (HIV) causes the disease known as AIDS (Acquired Immune Deficiency Syndrome). This disease is transmitted via bodily fluids such as blood.

17. **D** Earthquakes can affect vast areas of land. For example, when the December 26, 2004, earthquake struck in Indonesia, the tsunami it produced killed people as far away as the other side of the Indian Ocean, along the coast of Africa.

18. **B** Asbestos fibers range in size from 0.1 to 10 μm in length, and their inhalation causes asbestosis. Asbestosis is a form of scarring that occurs in the lungs; workers suffering from asbestosis may develop a slow buildup of scar-like tissue in the lungs and the membrane that surrounds the lungs.

19. **A** Radon is a radioactive gas that results from the radioactive decay of uranium. Since uranium is naturally occurring in some soils, radon can enter homes through cracks or windows in basement walls. This radioactive gas increases the risk of developing lung cancer.

20. **C** The parasite that causes malaria (Plasmodium sps.) is carried by mosquitoes (Anopheles sps.). This parasite is passed from person to person when mosquitoes bite one person, and then another.

21. **B** In grasslands, during the warm, wet season, grass grows rapidly, but during the cold winters and the annual dry periods, there is little growth. These long stretches of virtually no growth allow for the accumulation of biomass, which decays and creates a significant organic layer in the soil. Fires are typical in grasslands because of the dry grasses, and when they occur they help to break down plant material in this layer and release nutrients.

22. **D** Pine needles are full of acid material, such as pine tar and tannic acids. Large amounts of these needles create a soil that is slightly acidic.

23. **A** Since there is little plant growth in desert areas, there is little organic material in the soil; because these areas are arid, the soil is dry.

24. **C** In tropical rain forests, the decomposition of any organic material on the forest floor is very rapid due to moisture. This rapid decay produces acids that enter the soil, and the growing plants quickly absorb the products of decay.

25. **E** In deciduous forests, much growth occurs during the summer months, as fallen organic materials decay on the forest floor. During the winter, after the leaves fall, there is little decomposition of the leaves; this allows organic matter to accumulate.

26. **B** Nitrogen dioxide combines with atmospheric water to create nitric acid.

27. **E** Dust and soot are particles; with sizes ranging from 0.001 μm to 100.0 μm. They are both small enough to be suspended in the air.

28. **D** Ozone is one component of photochemical smog, along with aldehydes, PANs, and nitric acid.

29. **C** Sulfur dioxide combines with water to form sulfuric acid in the reaction

$$2SO_2 + O_2 \rightarrow SO_3, \text{ and then } SO_3 + H_2O \rightarrow H_2SO_4$$

30. **A** Carbon monoxide adheres to the blood's hemoglobin much more readily than oxygen. The CO stays bound to the hemoglobin and prevents hemoglobin from picking up oxygen from the lungs and transporting it to the tissues.

EXPLANATIONS FOR SECTION I—PART B

31. **E** Waste water treatment plants use a three-step process in removing waste from water. The main goals of sewage treatment are to remove the solid waste and reduce the biological oxygen demand (BOD). This BOD is a measure of how much organic material is in the water and how much bacteria can live in the water. The first stage of waste water treatment is the mechanical removal of solid objects, and screens of various sizes are used to perform this task. In the second stage, the water is sprayed on a bed of rocks that harbor billions of bacteria. The bacteria break down the organic molecules and lower the BOD. After the remaining liquid is treated, chlorine is put into the water to control bacteria populations.

32. **B** Insect larvae are the most, and most quickly, vulnerable to changes in abiotic factors like thermal pollution. As is true with the young of most species, the larval forms constitute the most vulnerable stage of an insect's life cycle. Species whose populations rise and fall with changes in abiotic factors are called indicator species. The presence of an indicator species in a certain ecosystem tells ecologists that the stream water is ideal to live in—if a fragile indicator species inhabits the region, then it must be safe for all species. If the indicator species is not there, then the water is not optimally clean.

33. **D** The LD_{50} is the value at which 50 percent of a population dies. If you read the graph, you can see that that value falls closest to the solution with the pH of 4.0. LD_{50} values are important because they help us understand the health risks of certain materials. As you might imagine, if a chemical is very toxic, it has a very low LD_{50}. For example, the nerve agent VX has an LD_{50} of 0.14 mg/kg body weight! It is extremely harmful to humans: This LD_{50} value tells us that it only takes 10.2 mg of nerve agent VX to kill half of the test population of 155-pound males! For the test, just remember that the smaller the LD_{50} number, the more toxic a chemical is.

34. **E** As you can see from the table, in the solution that had a pH of 1, almost no fish eggs hatched, so (E) is the correct answer. This graph analysis question asks you to understand that the size of the bar represents the percent surviving. You will almost certainly be asked to answer questions on this exam that include bar graphs, so if you need practice reading them, get it!

35. **B** The purpose of this test was to observe how many fish would hatch at different pH values, plain and simple. If you chose any other answer, then you might have been reading too much into the experiment. Remember not to make any inferences—you must answer the question only on the basis of the information you're given. In this experiment, the percent surviving is the dependent variable and the pH is the

independent variable; the data collected relates to hatching survival rates. pH is one of the most important abiotic factors for aquatic organisms. If the pH value shifts out of the ideal range, then the essential proteins in the cells that constitute fish eggs can denature. If that happens, the young will not survive.

36. **D** Only (D) correctly defines the meaning of pH. pH is the measure of the number of hydrogen ions in a solution. Chemically speaking, it is the negative logarithm (base 10) of the hydrogen ion concentration. So, if a solution has 10^{-7} hydrogen ions in it, then $-\log 10^{-7} = -(-7) = 7$. The pH of the solution is 7; the solution is neutral. The pH scale runs from 0–14; acidic solutions have a pH less than 7, while basic solutions have a pH greater than 7. Logarithms are used to keep the numbers simple, because the range of hydrogen ion concentrations can be as high as 100 trillion!

37. **A** The Comprehensive Environmental Response, Compensation, and Liability Act (CERCLA) was a law that was passed to establish the "Superfund." This law created a tax on the chemical and petroleum industries and provided broad federal authority to respond directly to releases or threatened releases of hazardous substances that may endanger public health or the environment. (B) is incorrect: RCRA established a system for managing non-hazardous and hazardous solid wastes in an environmentally sound manner. Specifically, it provides for the management of hazardous wastes from their point of origin to their point of final disposal (i.e., "cradle to grave"). RCRA also promotes resource recovery and waste minimization. (C) is responsible for air pollution control, and (D) controls water pollution standards. Finally, (E) mandated that environmental impact studies be done before major construction projects were started.

38. **A** In fact, increased global temperatures are responsible for the phenomena in most of the other answer choices! Global warming is more formally referred to as tropospheric warming, by the way. You learned that the atmosphere contains much H_2O vapor as well as trace amounts of CO_2, CH_4, and NO_2. These gases reflect infrared radiation that is, itself, reflected off the earth's surface as it is warmed by the sun. This process of tropospheric warming is called the "greenhouse effect," and it is a natural phenomenon. However, as humans burn huge amounts of fossil fuels, the excess CO_2 and NO_2 enter the atmosphere. These gases increase the heat-storing capacity of the atmosphere, which causes average global temperatures to increase. This, in turn, can cause later fall frosts and the northward migration of tree and plant species.

39. **D** This is an easy one. All of the answer choices are "positive" answer choices—which you would not think would cause high infant mortality—except (D). In fact, (D) is one strong indicator of countries that exhibit high infant mortality. Generally, high infant mortality rates occur in nations where the Gross Domestic Income (GDI) is low (less than $4,000). These nations cannot offer infant support services such as medicine, clean drinking water, sewage removal, and food. (A), (B), (C), and (E) all occur in countries that have low infant mortality rates.

40. **B** Composting is a process in which organic waste is converted into a rich soil conditioner called compost. Composting usually starts with household scraps made from plant materials, such as coffee grinds, vegetable waste, or anything derived from plants. Meat and dairy products are normally not used because they create very foul odors when they decompose, and attract undesirable animals to the compost bin. After the plant matter scraps are broken down, the material is added to soil to fertilize it and to help the soil retain moisture.

41. **C** (C) is the only false statement among the answer choices. This question requires that you be familiar with the first two laws of thermodynamics. The first law states that energy cannot be created or destroyed. That is, the total energy of a system will always remain the same. The second law states that at each

energy transformation, some energy is lost to the surroundings in an unusable form (usually as heat). For example, as a flashlight runs using battery power, some of the energy from the batteries goes toward warming the lightbulb, which is not useful work.

42. **C** The area of the Gulf of Mexico that extends from just south of the Mississippi River mouth to Texas receives nutrient-rich water from both the Mississippi and Atchafalaya Rivers. The water from these rivers carries runoff fertilizers from farms that lie in or near the watersheds for the rivers. These nutrients promote a rapid rise in the algae population in the Gulf waters, which in turn causes eutrophication. When the algal cells die, bacteria living in the water decompose them, or break them down; these decomposers use oxygen for metabolism, and eventually the water becomes extremely oxygen-poor. Low levels of oxygen in water often make these bodies of water uninhabitable for other aquatic organisms.

43. **A** It is a positive effect, while all of the other answer choices cite negative effects of genetically modified foods. Genetically modified foods (GMF's) are crop plants to which genes from other species have been added. For example, the "Flavorsavor" tomato contains select genes from fish! These genes were inserted into the tomato's genes so that the tomatoes will not freeze in a frost. Positive aspects for GMF's include increased yields; increased resistance to pests; and the potential to grow in habitats that previously did not support the plants.

44. **E** "Energy usefulness" is defined as how helpful an energy source is to humans. A source is more useful if it costs little to harvest, transport, and use. Useful energy is most commonly in a concentrated form. For example, coal is very highly useful, compared to wind energy; (A), (B), (C), and (D) are all examples of low-quality energy and matter.

45. **B** Review Chapter 4 if you have trouble remembering the basics of the geologic time scale! You will not be expected to memorize this chart, but you should be familiar with the most recent era, and the most talked-about ones. (B) correctly starts with the earliest era, the Precambrian (600 million years ago), then the Paleozoic (500 to 250 million years ago), then the Mesozoic (250 to 65 million years ago), and finally, the Cenozoic (65 million years ago to today)!

46. **C** Although the United States holds only 4.6 percent of the world's population, it produces 33 percent of the world's solid waste.

47. **B** A conservation easement is a voluntary agreement that allows a landowner to limit the type or amount of development on their property, while retaining private ownership of the land. The easement is signed by the landowner, who is the easement donor, and the Conservancy, who is the party receiving the easement. The Conservancy accepts the easement with understanding that it must enforce the terms of the easement in perpetuity. After the easement is signed, it is recorded with the County Register of Deeds and applies to all future owners of the land. In conservation easements, the owner usually receives a tax break for the promise that they will not build on that land.

48. **E** (E) is the defining policy statement of the CWA. This law governs the discharge of pollutants into the waters of the United States. It gives the Environmental Protection Agency the authority to implement pollution control programs and set wastewater standards for industry. The Clean Water Act also continues to set water quality standards for all contaminants in surface waters. The act makes it unlawful for any person to discharge any pollutant from a point source into navigable waters unless a permit was obtained under its provisions. It also funded the construction of sewage treatment plants under the construction grants program, and recognized the need to address the critical problems posed by non-point source pollution.

49. **A** Evolution is the change in the genetic makeup of a population over time. For this exam, the key is to remember that individuals do not evolve, and that only populations can evolve! (B) is not a good answer because a mutation is a change in DNA. (C) is also incorrect—evolution can occur in ways other than natural selection. (D) is wrong because it describes the movement of a population from one place to another; and finally, we can eliminate (E) because it refers to an increase, or magnification, of toxins in a food chain.

50. **A** This question requires you to understand the factors that determine the number of offspring that women are expected to have. Generally speaking, prosperity, high income, and a high level of education are all factors that lower a woman's total fertility rate (the number of offspring she will have). Statistically, the less wealthy and less educated a woman is, the higher her expected total fertility rate will be. (B) through (E) all list factors that lower the total fertility rate, so the correct answer is (A).

51. **A** (A) is a pretty good working definition of carrying capacity. Carrying capacity is denoted by "K," and refers to the number of individuals of a particular species that a certain geographic region can sustain, considering all of its available resources. (B) refers to the transition area between two habitats. (C) is called the doubling time; it's the segment of a population graph at which the population is doubling very quickly. (D) is a term that describes how populations evolve, and (E) defines all of the interacting species in an ecosystem.

52. **B** As chlorine atoms (from CFCs) catalyze the breakdown of ozone molecules in the stratosphere, more ultraviolet light is expected to reach the earth's surface. These UV rays will break down DNA molecules in human skin cells, which can lead to cancers, cataracts, and other illness in humans. Of course, other animals and plants can also be affected. While you may have been tempted to choose (A), (B) is a better option because it is more specific—(A) is much too general to be correct. (C), (D), and (E) are all results of global warming.

53. **B** Ozone is generated as a secondary pollutant in photochemical smog. Now this is an EXCEPT/NOT/LEAST question, so remember that you're looking for the incorrect answer! The exposure to tropospheric ozone can cause all of the symptoms given in (A), (C), (D), and (E). However, (B), lung cancer, is primarily caused by smoking cigarettes.

54. **C** The Net Primary Productivity (NPP) is the amount of energy that plants pass on to the community of herbivores in an ecosystem. It is calculated by taking the Gross Primary Productivity (the amount of sugar that the plants produce in photosynthesis) and subtracting from that the amount of energy the plants need for growth, maintenance, repair, and reproduction. NPP is measured in kilocalories per square meter per year (kcal/m2/y). (A) is incorrect because it denotes the amount of energy available for photosynthesis. (B) is no good; it merely states that energy cannot be created or destroyed. (D) can also be eliminated; it states that at each energy transfer, some energy is lost to the environment. Finally, (E) is incorrect because it deals with animals beyond herbivores.

55. **D** (D) is the only option that relates to the Second Law of Thermodynamics. Recall that the second law states that at each step of energy transformation, a large amount of energy is lost as heat to the environment. When we eat food, our bodies convert the energy in the food's chemical bonds to chemical bonds in our molecules. During this conversion, most energy is lost as heat (which is why our bodies are warm!).

56. **C** Acid deposition affects species by damaging habitat and by reducing or contaminating food sources through uptake of toxic levels of metals. Species such as amphibians, which require both aquatic and terrestrial environments, are most at risk. For example, in the acid-sensitive areas of eastern Canada, 16 of the 17 amphibian species have more than 50 percent of their ranges affected by acidic deposition. Monitoring amphibian populations may provide a biological indication of changes in acid deposition.

57. **C** Internal costs are all the costs that are incurred by the buyer and seller of the item in question. For example, the cost of gasoline and the purchase price of the car are both buyer costs. The costs of metal, plastics, employee salaries, and plant maintenance are internal costs of the seller. (C) is the correct answer because it is not an internal cost, but an external one; it is a cost paid by society. Health care costs that arise as a result of air pollution and the cost of building roads are both external costs.

58. **A** Scrubbers are devices that are installed in smoke stacks in order to reduce the amount of pollutants that rise up the stack. They separate out the pollutants in different ways: by trapping dust in filter material; using cyclonic air motion to concentrate dust in one area; or spraying water in the stack to trap both dust and gases that are water-soluble. (B) is simply not the best answer; (C) occurs before smoke is produced; (D) is not accomplished by scrubbers; and (E) is simply untrue—in fact, the amount of ash is greater when scrubbers are in place because scrubbers trap materials.

59. **D** After ore is mined, the unusable parts that are placed in piles are called tailings. This is a question that you'd either know the answer to or you wouldn't. Remember that if you come to a question like this and have absolutely no idea which choice is correct, pick your favorite letter and move on. Or circle it and come back to it in the next pass—maybe you'll remember the answer if you give your subconscious mind a chance to do its thing. Let's look at the other answer choices: (A), overburden, is the rock that's removed the ground in the surface mining of coal, so it's wrong. We made up (B), we hope you didn't choose it! (C), leachate, is the liquid that percolates through a mine or landfill, and (E), reclamation, is the process of restoring land to its former status after mining.

60. **D** Answer (D), "a walk in the woods" is an aesthetic value and not an economic good. An economic good is a physical object or service that has value to people and can be sold for a non-negative price in the marketplace.

61. **E** The fossil fuel coal is sedimentary rock that's derived from decayed and fossilized plant and animal materials. During the fossilization process, sulfur is incorporated into the rocks; this sulfur comes from bacteria that perform the decomposition of the material, as well as the decaying material itself. None of the other fuels contains as much of the element sulfur (S).

62. **C** (A), (B), (D), and (E) all have economic value, so they are all types of economic resources. Economic resources are reported as Gross National Income (GNI, formerly known as Gross National Product GNP). The GNI is the sum of all goods and services produced within a country, plus the value of its international trade. Know this for the exam!

63. **B** (B) is the best definition of a biological reserve. These reserves are areas of habitat that are crucial to the survival of species. Strictly speaking, a biological reserve is an area of land and/or water designated as having protected status for purposes of preserving certain biological features. Reserves are managed primarily to safeguard these features and provide opportunities for research into the problems underlying the management of natural sites and of vegetation and animal populations. Regulations are normally imposed on these areas that control public access and disturbance. All other options would lower the biodiversity of an area.

64. **D** As we reviewed in the Population chapter, China has the most people. (A) Japan, has about 128 million people; (B), the United States, has about 318 million people; (C), Brazil, has about 202 million people; (D), The People's Republic of China, has about 1.3 billion people; and (E), Russia, has only about 142 million people.

65. **E** Recall that the hydrologic cycle describes the movement of water through the biosphere. As the sun warms the water and the earth, water evaporates and forms water vapor. (E) is the only answer choice listed that requires solar energy—in order for water to move from the liquid state to the gaseous state, energy must be expended.

66. **B** (B) is the definition of full cost pricing: Full cost includes the environmental effects. All the other options are part of the external costs paid by the buyer or seller.

67. **A** The correct definition of parts per million is given in (A)—10 parts per million means there are 10 molecules of the substance in a total of 1 million molecules of the total solution. Remember that concentrations measure the amount of the solute in the solvent. An easy way to remember this is: The solvent always represents the larger volume, while the solute is present in lesser amounts. For example, when you make a glass of chocolate milk, the milk is the solvent (it has the larger volume) and the syrup is the solute (smaller volume).

68. **D** In the postindustrial state, the population has reached its carrying capacity and will exhibit a zero growth rate, or even drop below a zero growth rate.

69. **A** Methane is a useful biofuel that is produced from the decomposition of organic molecules in a landfill. This gas can be trapped, purified, and used to power vehicles. All of the other effects that are listed are undesirable results of the existence of old landfills.

70. **C** The CITES treaty was created in 1975, but was first discussed in the late 1960s when people recognized that there was no way to internationally protect endangered or threatened species. CITES is international in scope and is designed to halt the killing of endangered or threatened species for food, collection, or medicinal purposes by penalizing those who collect, trade, or buy those species.

71. **E** (E) uses the correct formula, which is (birth rate – death rate) ÷ 10. So, using the numbers in the problem, the correct calculation is $(12 - 9) \div 10 = 0.3$ percent.

72. **A** (A) is the most logical consequence of increased ocean levels. The increase is due to two main factors: The first factor is that as water warms, it expands. This is known as thermal expansion. Secondly, as the average global temperature increases, glaciers and other ice formations will melt; this will increase the volume of the world's oceans. All of the answer choices represent events that could be caused by an increase in the volume of the world's oceans.

73. **D** Compost is formed from decaying plants and other organic material. In the composting process, both bacteria and fungi decompose large organic molecules into smaller molecules. Compost can then be used to fortify and condition soil. While it might have appeared tempting, (C) is too general an answer, because all living organisms undergo respiration.

74. **E** Non-point pollution sources are defined as pollution that comes from a broad, ill-defined area, such as a large area of farmland. Since presumably, most farmers in the area fertilize their crops, you cannot point to one specific area as the only source. (A) through (D) are all point sources of pollution.

75. **D** (D) represents the standard definition of GDP. It is calculated by totaling the value of all the goods and services in the country for a specific time period, and is often considered as a measure of the standard of living in a country.

76. **E**   (E) is the correct answer; all other answer choices are open surface mining practices.

77. **B**   The total fertility rate in a country represents the average number of children a woman will have during her reproductive lifetime (between ages 15 to 45). The total fertility rate is affected by many factors, including level of education, culture, and the country's standard of living.

78. **B**   (B), or the fumes produced from the burning of wastes, is the only process that does not directly contribute to the contamination of the surrounding soil. When the EPA started to clean up the wastes from Love Canal, they had to deal with wastes that contaminated the surrounding land by all the other methods listed in the answer choices.

79. **C**   (C) provides the correct definition of an atom; atoms are the basic building blocks of compounds. Mixtures and compounds are both made up of many different atoms, and isotopes are merely different "versions" of the same atoms. For example, the isotope of carbon C-14 has the same chemical properties as the isotope of carbon known as C-12. Electrons are a component of an atom, so that is not the best answer.

80. **C**   The bicarbonate ion (HCO_3-) is produced when atmospheric carbon dioxide reacts with ocean water. Bicarbonate ions act as a buffer in seawater and allows the pH to be relatively stable, ranging between 7.5 and 8.5. (A), phosphoric acid, contains no carbon atoms. Neither methane, glucose, nor carbon monoxide exists in large quantities in water.

81. **B**   (B) is correct because areas rich in limestone or other basic minerals can neutralize the acid deposition products. Limestone is composed mostly of calcium carbonate ($CaCO_3$). This dissolves in water to form bicarbonate ions (HCO_3-), which acts as the buffering agent.

82. **C**   (C) is the only area that does not store significant amounts of phosphorus. Phosphorus is rarely present in gases. Additionally, it is relatively insoluble in water. It is found most frequently in rocks and minerals.

83. **B**   The current hypothesis for the formation of large volumes of oxygen in the atmosphere is called the autotrophic hypothesis. In it, the early atmosphere had almost no O_2 (g). It was not until the evolution of photosynthesis that the levels of O_2 increased. As for the other answer choices, volcanic outgas mostly consists of oxides of carbon, nitrogen, and hydrogen sulfide; meteorites could not carry gases effectively; animals produce CO_2 not O_2; and like volcanoes, bubbling geysers outgas oxides of nitrogen, oxygen, and hydrogen sulfide.

84. **E**   "Animal studies" refers to tests conducted on animals in laboratories, under controlled conditions. For example, this type of study is used to determine the dosage of a chemical that will kill 50 percent of a test population (LD_{50}). Epidemiological studies are done by examining human populations exposed to certain risks (for example, smoking). Statistical probabilities are mathematical studies that attempt to calculate the probability of an event occurring.

85. **A**   This is a point source, one dose of the pollutant. You can determine this because the BOD drops off, meaning that the pollutant is being diluted by unpolluted water. (B) is incorrect because it deals with temperatures in the atmosphere. (C) is a secondary air pollutant. (D) is a class of pollutants that result from the mixing of combustion products in the atmosphere. (E), deep wells, are locations where toxic materials are stored deep underground. It is unlikely that they would pollute the stream.

86. **B** This graph reading question tests your ability to make estimates from a graph. Look carefully at the axes and scale. It is helpful to use your pencil to draw a line from the 12-mile mark in the x-axis up to the line, and then another line from that point to the *y*-axis. You should be able to estimate 220 ppm (parts per million). (A) is too high an estimate. (C), (D), and (E) are estimates that are too low.

87. **E** The BOD, or Biological Oxygen Demand, test determines the rate at which microorganisms take oxygen out of the water. This test is an estimate of the amount of biodegradable organic matter in the water and is an indirect indicator of water quality. (A) is the turbidity test—the amount of particles that scatter light in the water. (B), nitrates, are important plant nutrients and are measured by tests that can quantitatively determine the concentration of the chemical ions. (C), the velocity test, measures the rate of water movement. (D), coliform bacteria, live in the intestinal tracts of animals and are an indication that fecal matter is in the water. This is not a BOD test because the coliform test is a count of the number of living cells, not a measure of their biological activity.

88. **A** You should recognize that the three organisms live in conditions of very low oxygen content. Looking carefully at the graph, you will see that the BOD is highest from 0 to 5 miles. Because the BOD is highest here, the amount of oxygen in the water is the lowest. Make sense? Remember, high demand means low amounts of oxygen in the water! (B), (C), (D), and (E) are in areas where the BOD is low—so there is more oxygen in the water. The three listed organisms do not live in areas of high oxygen content.

89. **B** (B) is a correct definition of a riparian zone. It is an ecotone between a river or stream and land. Riparian zones act as a buffer area, absorbing excess water and pollutants that travel from the land to the water. (A) defines rangeland, (C) is the role of a tree line; (D) is the understory of a forest; (E) defines headwaters.

90. **D** This is the only answer that is a disadvantage. (A), (B), (C), and (E) are all advantages of fish farming. When a fish (or any animal) population is dense, parasites and diseases can flourish.

91. **E** (E) is the only "environmental wisdom" viewpoint. All the rest are other planet management view points.

92. **A** There is a little saying that might help you remember which form of nitrogen is utilized by plants. A mnemonic is: "The plants ate the nitrate." Notice that nitrate has the same last three letters as "ate." (B) and (C) are parts of the nitrogen cycle, but are not absorbed by plants. (D), guano, is nitrogen-rich bird and bat droppings. (E) is not a nitrogen-containing compound.

93. **D** Thermal inversions occur when a warm layer of air prevents polluted air from rising up over a city. Normally, cool upper air allows the warm air to rise and the pollutants are dispersed along with the warm air. When an inversion occurs, the upper air is warmer and the polluted air cannot rise. This traps the pollution close to the earth's surface. (A) and (C) both deal with water; thermal inversions are atmospheric phenomena. (B) is the pattern in a normal atmosphere; (E) is a pattern that would disperse pollution.

94. **D** (D) is the correct answer, based on the principles of an environmentally sustainable economy. This type of economy is defined as a low-waste society that uses conservation, recycling, and reduction to keep its use of nonrenewable resources at sustainable levels. A sustainable society means that all practices will maintain society for foreseeable generations.

95. **C** Chlorofluorocarbons (CFCs) contain the element chlorine, which catalyzes the breakdown of ozone in the stratosphere. The energy for this reaction comes from the sun's ultraviolet radiation. It is estimated that one atom of chlorine will break down 100,000 molecules of ozone in the time it's in the atmosphere. (A) is a normal component of the atmosphere; (B) is a greenhouse gas; (D) and (E) cause acid deposition.

96. **C** (C) is the only series in which each number is double the one before; this is the definition of exponential growth.

97. **C** The ice cap that lies over much of Greenland is more than two miles thick, and has trapped gases from the atmosphere from hundreds of thousands of years ago. Other samples—(B), (D), and (E)—do not list an actual sample; methane is produced in the oil-making process, among other places.

98. **D** Bacteria are vital for the nitrogen cycle. They convert nitrogen from one form to another, and eventually into nitrates, which plants can use. (A), (B), (C), and (E) are not dependent on bacteria and fungi in the soil.

99. **D** By definition, trace elements are needed in small amounts. (B), (C), and (E) are typical components of fertilizers and are needed in large quantities; (A) is supplied by the atmosphere in large quantities.

100. **D** (D) defines the beliefs of the environmental justice movement. (A) is a system of beliefs in which people can learn to live in natural harmony with the planet. Those who have viewpoint (B) believe that humans are the best planet managers and that our technology and understanding of systems will help us make the correct decision; those who are ecofeminists are concerned with the relationship of women to the earth and male-dominated policies; and sustainability is the point of view that all people must live in ways that allow the earth to be sustainable.

FREE-RESPONSE EXPLANATIONS

Remember! You must write your responses in paragraph form!

Question 1

According to the United States Energy Information Administration, the consumption of natural gas by the United States increases at 8 percent per year. It receives its supplies from a variety of international and domestic locations. Natural gas is used in the home, industry, and in the generation of power.

(a) Calculate the approximate number of years it would take to double the consumption of natural gas in the United States. Show all work.

Apply the rule of 70 to calculate the doubling time. The formula is: $\frac{70}{8} = 8.75$. In other words, it would take about nine years to double the consumption. (Either an exact or rounded figure would be considered correct.)
(2 points maximum—1 point for the correct setup, 1 point for the answer)

(b) Describe one method by which natural gas is recovered and transported.

Natural gas is recovered through a series of pipes and valves fitted over a drilled hole. These are often associated with oil wells, coal beds, or natural gas pockets. The gas can then be transmitted through piping or cooled and compressed into Liquefied Natural Gas (LNG) and transported by specially fitted trucks, trains, or ships. Some natural gas is collected as a product of the anaerobic decay of organic material in landfills.
(2 points maximum—1 for a correct description of a collection method, 1 point for a correct transport method)

(c) Describe two benefits to the environment that would occur if the United States switched from coal to natural gas-fired electric power generation.

The environment would benefit through a reduction of the following:

- Less habitat destruction from the mining of coal

- Less acid mine drainage due to fewer mines being dug

- Less aesthetic damage done to the environment because fewer mines are dug

- Reduced SO_2 emissions, which will reduce the levels of acid deposition

- Reduced CO_2 emissions, which will reduce the levels of climate-changing gases

- Reduced soot emissions, which will reduce the formation of industrial (gray) smog

- Less fly ash produced

- Less mercury or radioactive materials released by the burning of coal

- Longer boiler life because gas produces fewer products that harm the boilers

- Less cost in building new generation plants, as gas combustion produces fewer dangerous by-products

(2 points maximum—1 point for benefit, 1 point for description. Remember, you only receive credit for your first two answers)

(d) Some people advocate increasing the use of coal versus natural gas for the production of e[lectricity]. Explain one argument that the proponents of coal might use to justify their position.

Coal-use proponents could make the following arguments:

- There is a larger supply of domestic coal than natural gas, so the United States will be less depend[ent on] foreign energy sources.

- Coal is safer to use. In an accident, natural gas can explode, harming people or facilities.

- The infrastructure for transporting coal already is in place. We would have to do little to increas[e the] production and transportation of coal.

- It might be cheaper to refit existing coal plants with scrubbers to clean up toxic emissions than to bui[ld a] new, cleaner-burning gas plant.

(2 points maximum—1 point for argument, 1 point for description)

(e) Possible answers:

Energy Source	Drawbacks
Nuclear	• Safety concerns (such as potential for radiation leaks) • Waste disposal concerns (such as radioactive or toxic waste disposal) • Raw materials storage (safe storage of radioactive materials) • Thermal pollution (reactor cooling fluids are very hot)
Hydroelectric	• Water loss via evaporation at dam site • Dam may hinder fish and other aquatic migration • Silting • Habitat destruction/alteration (via flooding upstream, or alteration of water quality downstream) • Potential for displacing populations or wildlife living in dam's flood zone • Potential for seismic activity below reservoir • Potential for plants killed by flooding to produce methane gas (a greenhouse gas)
Solar	• Initial costs are high • Material for solar panels uses hydrocarbon fuels • Solar panels lack aesthetic appeal • Disposal of toxic chemicals (from rechargeable batteries)
Wind	• Initial costs are high • Turbines lack aesthetic appeal • Sound and vibration produced is annoying • Turbines may affect bird migration patterns
Geothermal	• Geographical limitations (there are only certain places in the world where geothermal energy can be harvested) • Thermal pollution • Toxic chemicals (potential for releasing H_2S in environment)

(2 points maximum—1 point for listing energy source, 1 point for listing related drawback)

Question 2

The map below shows two cities: City X and City Z, separated by several kilometers. Students from a high school in between the two cities studied soil pH values at the sites labeled A through E on the map. The results of the pH study are given in the following table:

Site	pH value
A	6.2
B	5.6
C	5.0
D	4.5
E	4.3

(a) Describe one point source for the pollution that caused the change in the soil's pH as shown. Include in the description a fuel that could create the pollution.

Point sources could include

- Electricity generation plants
- Industry sites
- Cars or trucks
- Homes

    The fuels involved would be coal or oil. (Natural gas and nuclear would not be accepted as correct.) *(2 points maximum—1 for the point source and 1 for the fuel)*

(b) Identify one primary and one secondary pollutant that can cause the change in the soil's pH. Describe the process that causes the change in the pH.

Primary pollutants are either SO_x or NO_x (saying just "sulfur" or "nitrogen" is not correct, as they do not directly produce the acid reaction). Also, suspended particulate matter can carry acidifying chemicals.

As the fuel is burned, SO_x or NO_x are produced; these enter the atmosphere and react with water vapor to create sulfuric acid or nitric acid. Farther downwind, these chemicals fall as wet or dry deposition and acidify the soil.
(4 points maximum—1 for identifying the primary pollutant and 3 for describing the process of acid formation and deposition)

(c) Describe one possible method to reduce the air pollutants that are causing the pH change.

Methods of reducing the air pollutants that are causing the pH change include

- Reducing factory emissions of SO_x or NO_x by using scrubbers to remove chemicals from smoke
- Using cleaner-burning fuels (natural gas, low-sulfur coal, etc.)

- Reducing demand for electricity, which lowers production

- Adding catalytic converters to lower pollution from cars and trucks

  (2 points maximum—1 for the method and 1 for the correct description)

  (d) Describe one provision of the Clean Air Act of 1990 that could be used to control and reduce the emissions.

- Provisions of the Clean Air Act of 1990 that could be used to control and reduce the emissions include

- National Ambient Air Quality standards—a set of maximum permissible levels for pollutants

- Emissions trading policy—each year certain factories are given "rights" or "credits" to release set amounts of pollutants. These credits can be bought, sold, or traded to reduce a company's liability under the CAA.

- National emission standards for hundreds of toxic pollutants (e.g., mercury)

  (2 points maximum—1 for the policy and 1 for the correct description of the policy)

Question 3

Question 3 referred to the editorial on page 301.

(a) Describe how the beetle might have become resistant to NOGrub. Assume that NOGrub had been applied to a population of beetles in another county.

In a sexually reproducing population, there is natural variation in the genetic makeup of the population. There are large numbers of beetles, so some of them have a genetic makeup that allows for their survival in an environment with NOGrub in it. When the pesticide is applied, the susceptible beetles die and the resistant ones survive. The survivors reproduce in a habitat of lessened competition, therefore their genes are passed along.
(3 points maximum—1 for the idea of natural selection, 1 point for selective survival, and 1 point for growth of resistant populations)

(b) Discuss two negative impacts of using chemical pesticides on the surrounding ecosystem.

Negative impacts of using chemical pesticides on the surrounding ecosystem include

- Pesticides can migrate to other habitats, thus harming other organisms

- Pesticides can kill beneficial insects, which act as predators of destructive insects

- Pesticides can combine with other chemicals and create other toxic chemicals

- Pesticides can be accidentally ingested by humans and cause illness

  (2 points maximum—1 for each correct impact)

(c) Describe two other methods of controlling the beetles without resorting to human-made chemical pesticides.

Some methods of controlling the beetles without resorting to human-made chemical pesticides are

- Cultivation practices that reduce insect populations—timing plantings to reduce insect populations
- Genetic engineering to make plants pest resistant—plants produce chemicals that kill insects
- Control insect reproduction through sterilized males or hormone control
- Sex attractants—chemicals produced by females (pheromones) that attract males
- Hormones that disrupt the normal development of insects from egg, to larvae, to adult
- Release of an insect's natural predators
  (2 points maximum—1 for each correct method)

(d) Possible answers:

(i) 1 point for showing an understanding of one of the following:

- IPM employs a combination of biological, chemical, and physical (any two of the three) methods for pest control.
- Goal of IPM is to curtail (or eliminate) pesticide use.
- IPM generally does not eradicate the pest population, but brings it to a tolerable level.

(ii) Explain one benefit and one difficulty in using IPM to control this outbreak.

Benefits of using integrated pest management	Difficulties in using integrated pest management
Lowers costs for pesticides	Need expert knowledge of insect life cycle
Reduces amounts of pesticides in environment	Takes time to implement management practices
Several weaknesses of insects can be exploited	Initial costs might be higher
Reduces use of fertilizers	Governments subsidize pesticide use
Improves crop yields	Training takes a long time
Lowers genetic resistance issues	Methods for one area might not apply in other areas

(2 points maximum—1 for a correct benefit, and 1 for a correct difficulty)

Question 4

Many endangered species live in areas where biodiversity has been degraded by human activities. Species such as the West Virginia spring salamander or the California condor live in areas where the impact of human activities has made these and other organisms very rare.

(a) Discuss two human activities that cause species to become endangered.

Activities include

- Habitat destruction—farming, logging, expansion of cities or villages

- Hunting and Poaching—using animal parts for food or medicinal purposes

- Introduction of alien species—these out-compete the native species for niches in the habitat

- Sale of exotic pets and plants—tropical fish, plants such as orchids, and rare birds command high prices from collectors

- Predator and pest control—chemicals used to control pests can harm native species

- Pollution—toxic chemicals can kill sensitive species or threaten their food supply

  (4 points maximum—2 for each activity, with 1 point for activity and 1 point for description)

  (b) Describe two reproductive strategy characteristics that make a species prone to extinction.

Reproductive strategy characteristics that make a species prone to extinction are

- Fewer, larger offspring—loss of an infant has greater effect on numbers

- High level of parental control—much time, energy, and resources given to care of young; habitat loss can lower levels of care

- Later age of reproduction—need more environmental resources to survive

- Larger adults—easier to hunt and poach

- Lower population growth rate—habitat changes affect a larger percent of the population

- Specialized niche—habitat destruction reduces available resources

  (4 points maximum—2 for each strategy, with 1 point for strategy and 1 point for description)

  (c) Describe one economic and one legislative action that can be used to save endangered species.

Economic

- Wildlife refuges

- Gene banks

- Zoos/aquariums

- Farms for endangered species

- Eco–tourism

Legislation

- Endangered Species Act

- Lacey Act

- Convention on International Trade of Endangered Species (CITES); list of 900 species that can not be traded internationally

  (2 points maximum—1 point for economic and 1 point for legislative action)

HOW TO SCORE PRACTICE TEST 1

Section I: Multiple-Choice

$$\underline{\hspace{3cm}} \times .9000 = \underline{\hspace{3cm}}$$

Number Correct   Weighted
(out of 100)   Section I Score
                 (Do not round)

Section II: Free Response

Question 1 $\underline{\hspace{3cm}} \times 1.5000 = \underline{\hspace{3cm}}$

                (out of 10)   (Do not round)

Question 2 $\underline{\hspace{3cm}} \times 1.5000 = \underline{\hspace{3cm}}$

                (out of 10)   (Do not round)

Question 3 $\underline{\hspace{3cm}} \times 1.5000 = \underline{\hspace{3cm}}$

                (out of 10)   (Do not round)

Question 4 $\underline{\hspace{3cm}} \times 1.5000 = \underline{\hspace{3cm}}$

                (out of 10)   (Do not round)

AP Score Conversion Chart Environmental Science

Composite Score Range	AP Score
98–150	5
78–97	4
66–77	3
55–65	2
0–54	1

Sum = $\underline{\hspace{3cm}}$

            Weighted
            Section II Score
            (Do not round)

Composite Score

$$\underline{\hspace{3cm}} + \underline{\hspace{3cm}} + \underline{\hspace{3cm}}$$

Weighted   Weighted   Composite Score
Section I Score   Section II Score   (Round to nearest
                                       whole number)

Chapter 16
Practice Test 2

AP® Environmental Science Exam

SECTION I: Multiple-Choice Questions

DO NOT OPEN THIS BOOKLET UNTIL YOU ARE TOLD TO DO SO.

Instructions

At a Glance

Total Time
1 hour and 30 minutes
Number of Questions
100
Percent of Total Grade
60%
Writing Instrument
Pencil required

Section I of this examination contains 100 multiple-choice questions. Fill in only the ovals for numbers 1 through 100 on your answer sheet.

Indicate all of your answers to the multiple-choice questions on the answer sheet. No credit will be given for anything written in this exam booklet, but you may use the booklet for notes or scratch work. After you have decided which of the suggested answers is best, completely fill in the corresponding oval on the answer sheet. Give only one answer to each question. If you change an answer, be sure that the previous mark is erased completely. Here is a sample question and answer.

Sample Question   Sample Answer

Chicago is a

(A) state
(B) city
(C) country
(D) continent
(E) village

Use your time effectively, working as quickly as you can without losing accuracy. Do not spend too much time on any one question. Go on to other questions and come back to the ones you have not answered if you have time. It is not expected that everyone will know the answers to all the multiple-choice questions.

About Guessing

Many candidates wonder whether or not to guess the answers to questions about which they are not certain. Multiple choice scores are based on the number of questions answered correctly. Points are not deducted for incorrect answers, and no points are awarded for unanswered questions. Because points are not deducted for incorrect answers, you are encouraged to answer all multiple-choice questions. On any questions you do not know the answer to, you should eliminate as many choices as you can, and then select the best answer among the remaining choices.

GO ON TO THE NEXT PAGE.

ENVIRONMENTAL SCIENCE
Section I
Time—90 minutes
Part A

Directions: Each set of lettered choices below refers to the numbered questions or statements immediately following it. Select the one lettered choice that best answers each question or best fits each statement and then fill in the corresponding oval on the answer sheet. A choice may be used once, more than once, or not at all in each set.

Questions 1-5 refer to the following kinds of species interactions.

    (A) Parasitism
    (B) Mutualism
    (C) Competition
    (D) Predator/prey
    (E) Commensalism

1. Cheetahs and antelope

2. Mycorrhizae and plant roots

3. Humans and tapeworms

4. A hawk eats a mouse

5. Two birds fight over a nest

Questions 6-10 refer to the following kinds of federally owned lands.

    (A) National Forests
    (B) National Resource Lands
    (C) National Wildlife Refuges
    (D) National Parks
    (E) National Wilderness Preservation Areas

6. These lands are reserved for recreation and fishing, but not for commercial use.

7. These multiuse lands can be logged and fished.

8. These lands provide a domestic energy and mineral supply.

9. Motor vehicles are banned from these hunting and fishing areas.

10. These habitats provide protected breeding areas for many animals.

Questions 11-15 refer to the following environmental disasters.

    (A) Chernobyl, Ukraine
    (B) Love Canal, New York
    (C) Bhopal, India
    (D) London, England
    (E) Chesapeake Bay, Maryland

11. When this chemical plant exploded, thousands of people died from inhaling pesticides.

12. Chemicals leaking out of a dump caused leukemia in local children.

13. The burning of coal produced smog so thick that many people died of respiratory illnesses.

14. This estuary is eutrophied, and many plant and animal species are now unable to live there.

15. An explosion in a nuclear reactor spread radioactive materials over thousands of hectares in this region.

GO ON TO THE NEXT PAGE.

Questions 16-20 refer to the following waste disposal methods.

(A) Landfill
(B) Incineration
(C) Chemical treatment
(D) Biological treatment
(E) Discharge into sewers and rivers

16. This method can produce leachate.

17. This method reduces the acidity of a solution.

18. This method can release toxic gases.

19. This method can cause high biological oxygen demand (BOD).

20. This method involves the use of bacteria and fungi.

Questions 21-25 refer to the following renewable energy sources.

(A) Solar energy
(B) Hydrogen fuel cells
(C) Tidal energy
(D) Geothermal energy
(E) Wind energy

21. This source can harm migratory birds.

22. Water and electricity are products of this source.

23. This source of energy relies on water flowing in and out of bays.

24. This source converts radiant energy into heat or electricity.

25. This source utilizes heat or steam from deep underground.

Questions 26-30 refer to the following forms of air pollution.

(A) O_3
(B) SO_2
(C) Pollen
(D) Soot
(E) Hydrocarbons

26. This gas originates from the combustion of coal.

27. Produced by plants, this can cause allergic reactions.

28. This can be destroyed by CFCs.

29. One source of this is spilled gasoline.

30. Small particles that can irritate both the eyes and lungs.

GO ON TO THE NEXT PAGE.

Part B

Directions: Each of the questions or incomplete statements below is followed by five suggested answers or completions. Select the one that is best in each case and then fill in the corresponding oval on the answer sheet.

31. Exposure to which of the following noises would cause the most damage to a person's hearing?

    (A) A vacuum cleaner
    (B) A chain saw
    (C) A factory
    (D) The firing of a rifle
    (E) A lawn mower

32. The phrase that best defines population density is

    (A) the number of individuals in a certain geographic area
    (B) the rate at which a population increases
    (C) the maximum number of individuals that a habitat can sustain
    (D) the time it takes for a population to increase to carrying capacity
    (E) one way that populations can be grouped in a certain geographic area

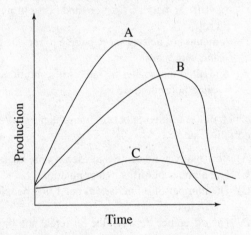

33. The graph above represents possible depletion curves of a nonrenewable resource. Curve C best describes the resource as it

    (A) has just been discovered and no technology exists to use the resource
    (B) is quickly used up and is not recycled
    (C) is newly discovered and in high demand
    (D) has expanding reserves and consumption is reduced
    (E) has some recycling and no reserves are being discovered

34. The shrinking of the Aral Sea and the ecological disaster that followed was mainly caused by

    (A) the diversion of the sea's two feeder rivers for agricultural use
    (B) withdrawing groundwater from the area
    (C) a major earthquake that hit the region
    (D) the loss of swamplands in the area
    (E) the massive use of pesticides

35. Which of the following is a negative impact of overfishing a particular species of edible fish?

    I. Loss of so many fish that there is no longer a breeding stock
    II. The removal of non-target species
    III. The reduction of other species that rely on the edible species as food.

    (A) I only
    (B) II only
    (C) I and II only
    (D) II and III only
    (E) I, II, and III

36. Which type of irrigation results in the greatest amount of water lost to evaporation?

    I. Flood irrigation
    II. Drip irrigation
    III. Low-pressure central pivot irrigation

    (A) I only
    (B) II only
    (C) III only
    (D) I and III only
    (E) II and III only

GO ON TO THE NEXT PAGE.

Questions 37-40 refer to the following graph.

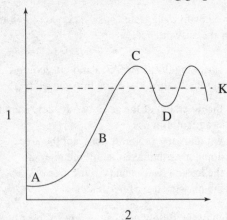

37. The population is growing at its highest rate at which letter?

    (A) A
    (B) K
    (C) D
    (D) B
    (E) C

38. Which of the following phrases best describes line K in the graph above?

    (A) The rate of population growth
    (B) The carrying capacity of the environment
    (C) The time it takes for the population to double in size
    (D) The number of individuals in the population at the beginning of the study
    (E) The birth rate of the population

39. The best label for the axis labeled 2 in the graph above is

    (A) time
    (B) population number
    (C) logistic growth rate
    (D) environmental resistance
    (E) birth rate

40. Which of the following statements is true concerning the events occurring at point D in the graph above?

    (A) Environmental resistance is high.
    (B) Environmental resistance is low.
    (C) The population will continue to fall.
    (D) The prey population will soon rise.
    (E) The population is above its carrying capacity.

41. The long-term storage of phosphorus and sulfur is in which of the following forms?

    (A) Bacteria
    (B) Rocks
    (C) Water
    (D) Plants
    (E) Atmosphere

42. The movement of sections of the earth's lithosphere is known as

    (A) mass depletion
    (B) plate tectonics
    (C) background extinction
    (D) migration
    (E) emigration

43. Which of the following best defines the Green Revolution?

    (A) An international effort to stop the construction of nuclear power plants
    (B) A group whose goal is to improve how nations affect the environment
    (C) Increasing the yield of farmland by using more fertilizer, better irrigation, and faster growing crops
    (D) A method of getting more people to recycle paper and cardboard
    (E) A method that makes a viable soil conditioner by using household waste

44. Which of the following best defines the phrase "infant mortality rate"?

    (A) How many children live in each square hectare
    (B) The number of births in a population
    (C) The number of infant deaths per 1,000 people aged zero to one
    (D) The difference between the birth rate and the death rate in a population
    (E) The total number of children in a population

45. The release of chemicals from underground storage tanks is most likely to pollute which of the following?

    (A) A landfill
    (B) The atmosphere
    (C) The ecotone
    (D) Aquifers
    (E) The hydrosphere

GO ON TO THE NEXT PAGE.

46. The United States Congress recently failed to ratify which of the following international agreements that is designed to control the release of carbon dioxide?

   (A) CITES agreement
   (B) Kyoto Protocol
   (C) Montreal Protocol
   (D) Clean Air Act
   (E) RCRA

47. Which of the following is the most sustainable way to ensure sufficient energy for the future?

   (A) Find more fossil fuels
   (B) Develop more effective solar power generators
   (C) Build more nuclear reactors
   (D) Reduce waste and inefficiency in electricity use and transmission
   (E) Increase the amount of electricity generated using natural gas

48. Which of the following best describes the goals of the CAFE standards?

   (A) Reduce pollution by coal fired power plants
   (B) Increase habitat diversity in certain areas
   (C) Improve the quality of air around cities
   (D) Protect certain endangered species
   (E) Improve the fuel efficiency of automobiles in the United States

49. Which of the following best describes the use of DDT?

   (A) It supplies needed nitrogen to plants
   (B) It kills weeds and unwanted plants
   (C) It decreases the amount of pollution from car exhaust
   (D) It is a chemical involved in photosynthesis
   (E) It is an insecticide

GO ON TO THE NEXT PAGE.

Questions 50-52 refer to the following world population graph.

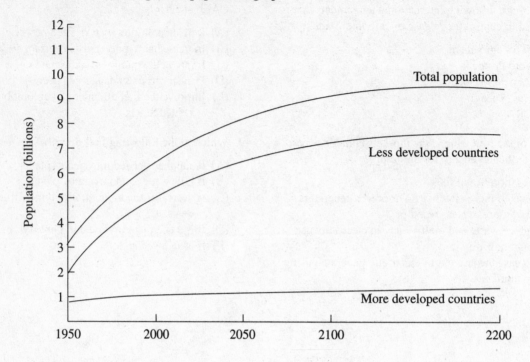

50. Which of the answers below best describes the change in the population of less developed countries between the years 1950 and 2050?

    (A) There will be an increase of approximately 1 billion.
    (B) There will be an increase of approximately 4 billion.
    (C) There will be an increase of approximately 10 billion.
    (D) There will be a decline of 1 billion.
    (E) There will be a decline of 6 billion.

51. According to the graph, the total population of the world is most likely to increase due to which of the following?

    (A) The rise in populations of developed countries
    (B) The continued emigration of people
    (C) The use of more fossil fuels
    (D) The rise in populations of developing countries
    (E) An increase in the amount of food produced in less arable land

52. According to the graph, in the year 2000 the less developed countries had a population that was how many times bigger than the developed countries?

    (A) 12 times
    (B) 8 times
    (C) 4 times
    (D) 1.5 times
    (E) 0.5 times

GO ON TO THE NEXT PAGE.

Fresh Water Usage, United States 2000

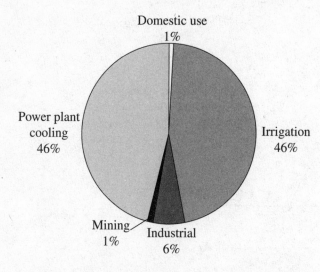

53. According to the diagram above, cooking, showering and using toilets accounted for approximately what percent of total water use?

(A) 92 percent
(B) 46 percent
(C) 40 percent
(D) 6 percent
(E) 1 percent

54. One result of increased troposphere temperatures that are observed today is

(A) an increase in skin cancers in people
(B) an increase in the average global sea level of 10 to 20 cm
(C) more radon seepage into people's homes
(D) a deeper permafrost in Arctic regions
(E) lakes and ponds remaining frozen longer into the spring

55. All of the following gases contribute to rising global air temperatures EXCEPT

(A) methane
(B) nitrogen dioxide
(C) ozone
(D) carbon dioxide
(E) water vapor

56. Which of the following is the root cause of habitat loss; especially in less developed nations?

(A) road building
(B) poverty
(C) conversion of forest to farmland
(D) capturing exotic animals for resale
(E) a warm, moist habitat

57. Per capita income can best be defined as

(A) how long the average person lives
(B) the concentration of people in a city
(C) the total amount of income in a country
(D) how much money each person makes
(E) how many people are worth more than one million dollars

58. The motion of tectonic plates accounts for most of Earth's

(A) CO_2 emissions
(B) formation of rivers
(C) change of seasons
(D) volcanic activity
(E) oil formations

59. "K" and "r" are used to describe which of the following aspects of populations?

(A) The place in a habitat where these organisms live
(B) The number of males and females in the population
(C) The number of predators each population has
(D) The reproductive tactics used by populations
(E) The time it takes a population to double

60. Convectional heating and cooling of the atmosphere transfers which of the following to other parts of the earth?

I. Heat
II. Moisture
III. Nutrients

(A) I only
(B) II only
(C) III only
(D) I and II only
(E) I, II, and III

GO ON TO THE NEXT PAGE.

Questions 61 and 62 refer to the following diagram.

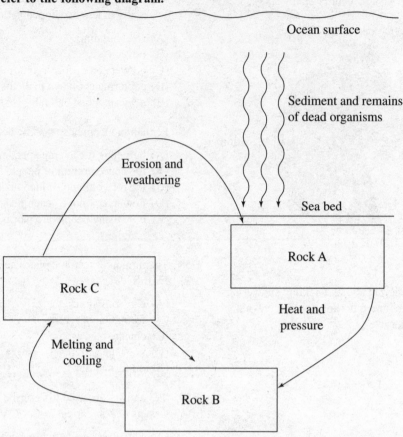

61. Which of the following gives the correct sequence in the rock cycle shown above?

    (A) Rock A—Metamorphic/Rock B—Igneous/
        Rock C—Sedimentary
    (B) Rock A—Sedimentary/Rock B—Metamorphic/
        Rock C—Igneous
    (C) Rock A—Igneous/Rock B—Sedimentary/
        Rock C—Metamorphic
    (D) Rock A—Metamorphic/Rock B—Sedimentary/
        Rock C—Igneous
    (E) Rock A—Sedimentary/Rock B—Igneous/
        Rock C—Metamorphic

62. Which of the following rock types would contain the greatest number of fossils?

    (A) Igneous rock only
    (B) Metamorphic rock only
    (C) Sedimentary rock only
    (D) Igneous and metamorphic rock
    (E) Sedimentary and igneous rock

GO ON TO THE NEXT PAGE.

63. The North Atlantic Current provides which of the following for Europe and North America?

    (A) Fish to feed predators such as killer whales
    (B) Warm water that moderates land temperatures
    (C) Large amounts of CO_2 to promote photosynthesis
    (D) Cold saltwater to help form icebergs
    (E) Mineral-rich waters to reduce depleted mineral reserves

64. Which of the following indoor air pollutants is composed of microscopic mineral fibers that can produce lung cancer in humans?

    (A) Nitrogen oxides
    (B) Carbon monoxide
    (C) Asbestos
    (D) Radon
    (E) Formaldehyde

65. Which of the following atoms is the primary catalyst in the destruction of ozone?

    (A) Fluorine
    (B) Oxygen
    (C) Mercury
    (D) Platinum
    (E) Chlorine

66. Transpiration is best defined as the

    (A) movement of water through aquifers
    (B) evaporation of water from an ocean
    (C) movement of water into the air from the leaves of plants
    (D) condensation of water into rain
    (E) heating of the atmosphere by the warmth of the earth

67. Nuclear reactors use which of the following to absorb neutrons in the reactor core?

    (A) Steam condenser
    (B) Control rods
    (C) Heat exchanger
    (D) Fuel rods
    (E) Turbines

68. Riparian areas are vital to the preservation of high-quality

    (A) mountain slopes
    (B) grazing land
    (C) feedlots
    (D) rivers and streams
    (E) ocean beaches

69. Which of the following fishing techniques is most damaging to ocean bottom ecosystems?

    (A) Trawling
    (B) The use of drift nets
    (C) Long lines
    (D) Purse-seine
    (E) Fish farming

70. Which of the following treaties is responsible for lower levels of CFC production world wide?

    (A) Montreal Protocol
    (B) Kyoto Protocol
    (C) Clean Air Act
    (D) The Rio Earth Summit of 1972
    (E) Comprehensive Environmental Response, Compensation and Liability Act

71. DDT is an insecticide sprayed to control insects. Years after it was introduced, DDT was found in large predatory birds, such as the osprey. Which of the following processes caused the DDT to be found in the osprey?

    I. Biomagnification
    II. Bioremediation
    III. Bioaccumulation

    (A) I only
    (B) II only
    (C) III only
    (D) I and III only
    (E) I, II, and III

72. Doing which of the following could most effectively reduce acid rain and acid deposition?

    (A) Reducing the use and waste of electricity
    (B) Making taller smoke stacks
    (C) Burning coal twice
    (D) Adding lime to acidified lakes
    (E) Moving power plants to desert areas

73. Which of the following forms of radiation is most harmful to humans?

    (A) Alpha
    (B) Gamma
    (C) Beta
    (D) Infrared
    (E) Radon

GO ON TO THE NEXT PAGE.

74. Which of the following ecosystems does NOT use solar energy as its ultimate energy source?

    (A) Pond
    (B) Deep-sea hydrothermal vent
    (C) Rain forest
    (D) Tundra
    (E) Coniferous forest

75. All of the following are true about CO_2 sequestering EXCEPT

    (A) it can be accomplished by pumping CO_2 into carbonated beverages
    (B) it can be accomplished by pumping CO_2 into crop lands
    (C) it can be accomplished by pumping CO_2 deep under the ocean floor
    (D) it can be accomplished by pumping CO_2 deep underground into dried up oil wells
    (E) it can be accomplished by pumping CO_2 into immature forests

76. Which of the following is NOT true concerning invasive species?

    (A) They can out-compete native species in a habitat.
    (B) They can reproduce more rapidly than native species.
    (C) They are highly specialized and have narrow niches.
    (D) They alter the biodiversity of the area they are invading.
    (E) They are introduced into a habitat and are not native.

77. The Second Law of Thermodynamics is best described by which of the following?

    (A) The amount of solar radiation going into an ecosystem is equal to the total amount of energy going out of that system.
    (B) The amount of carbon in the atmosphere has increased due to the combustion of fossil fuels.
    (C) As electricity is transmitted through wires, some of the power is lost to the environment as heat.
    (D) Wind-generated electricity has more power than electricity generated at a hydropower plant.
    (E) The amount of electricity used to light a lightbulb is less than the amount of light that the bulb produces.

78. Salinization is a process that makes soil less productive because it

    (A) removes essential soil nutrients
    (B) lowers the pH
    (C) increases the salt content
    (D) makes the soil waterlogged
    (E) produces larger soil particles

79. Which of the following is true about early-loss populations, such as fish?

    (A) The chances of an adult dying are about the same as a child dying.
    (B) The maturation process is slow.
    (C) The populations are close to the carrying capacity.
    (D) They are specialists.
    (E) Many individuals die at an early age.

GO ON TO THE NEXT PAGE.

Questions 80 and 81 refer to the illustration of succession below.

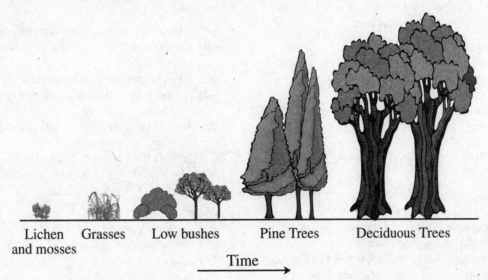

Lichen   Grasses   Low bushes   Pine Trees   Deciduous Trees
and mosses

Time →

80. A farmer stops farming a certain tract of land, and small bushes soon grow there. The land then progresses to the deciduous tree stage. This process is known as

(A) pioneer succession
(B) wetland succession
(C) secondary succession
(D) primary succession
(E) regressive succession

81. According to the diagram, low species diversity and small sized plants are characteristic of which stage of succession?

I. Late stage succession
II. Midstage succession
III. Early stage succession

(A) I only
(B) II only
(C) III only
(D) I and III only
(E) I, II, and III

GO ON TO THE NEXT PAGE.

82. In this method of land preservation, countries promise they will preserve important habitats in return for the forgiveness of loans to other countries.

    (A) Ecotourism
    (B) Establishment of biological reserves
    (C) Small-scale sustainable farming
    (D) Debt for nature swaps
    (E) Phase-out economic practices such as farm subsidies

83. Ozone depletion is occurring most rapidly in the earth's polar regions because

    (A) there are more researchers studying the problem at the poles than at the equator
    (B) the atmosphere is thicker at the poles, so ozone destruction is easier to observe
    (C) large amounts of chlorofluorocarbons (CFCs) can accumulate on ice crystals formed in the cold atmosphere
    (D) the upper atmosphere winds form a pattern of high and low pressure systems that can cause the destruction of ozone
    (E) the solar UV radiation is stronger at the poles, promoting the breakdown of ozone

84. Smaller forest fires are beneficial to forests for all of the following reasons EXCEPT

    (A) removal of competing plants
    (B) combustion of dried leaves or needles, which reduces the threat of large fires
    (C) burning the crowns of trees
    (D) germinating seeds of certain plant species
    (E) making burned matter available as a nutrient

85. What are the negative impacts of dams on ecosystems?

    I. Loss of silt in the river downstream from the dam
    II. Generation of low pollution electricity
    III. Loss of terrestrial biodiversity in areas surrounding the dam

    (A) I only
    (B) II only
    (C) III only
    (D) I and III only
    (E) I, II, and III

86. Which of the following best illustrates the process of evolution?

    (A) A parasite population becomes resistant to a drug
    (B) Rabbits can have brown fur in summer and white fur in winter
    (C) Frogs borrow deep into the mud during winter
    (D) A baby is born and has a different color hair than its parents
    (E) A squid changes color to hide from predators

87. Compounds such as water, can exist as a liquid, solid, or a gas. Changes between states of a substance are known as

    (A) phase changes
    (B) chemical changes
    (C) compound changes
    (D) nuclear reactions
    (E) aerobic respiration

88. The energy necessary to produce stratospheric ozone comes from which of the following?

    (A) Oxygen
    (B) Sunlight
    (C) Radioactive decay
    (D) Magma
    (E) Wind

89. Which of the following chemicals can cause lung irritation in the troposphere but is very helpful to humans in the stratosphere?

    (A) O_2
    (B) O_3
    (C) Chlorofluorocarbons
    (D) H_2SO_4
    (E) DDT

90. Which of the following best describes the differences between a primary pollutant and a secondary pollutant?

    (A) Primary pollutants rise up the smoke stack before secondary pollutants are formed.
    (B) Primary pollutants are formed from secondary pollutants interacting in the water.
    (C) Secondary pollutants are formed from primary pollutants interacting in the atmosphere.
    (D) Secondary pollutants are made by cars, while secondary pollutants are made by the burning of wood.
    (E) Secondary pollutants are directly created by the burning of coal and primary pollutants from the burning of oil.

GO ON TO THE NEXT PAGE.

91. An appliance operates at 120 volts and 10.0 amps for 1 hour. How many watt-hours does it use in that hour?

(A)   1.2 watt-hours
(B)   12 watt-hours
(C)   100 watt-hours
(D)   110 watt-hours
(E)   1200 watt-hours

92. Coal, oil, and natural gas were all formed as a result of

(A) the decay of organic matter
(B) the movement of magma in volcanoes
(C) sedimentary rock turning into metamorphic rock
(D) the radioactive decay occurring inside Earth
(E) the motion of the earth's core

93. All of the following are negative impacts of food production EXCEPT

(A) increased erosion
(B) air pollution from fossil fuels
(C) bioaccumulation of pesticides
(D) lower death rates
(E) loss of biodiversity

94. Soils found in mid-latitude grasslands would be most accurately described as having

(A) a high acid content with little organic matter
(B) a deep layer of humus and decayed plant material
(C) a layer of permafrost right below the O-horizon
(D) a high content of iron oxides and very little moisture
(E) a small amount of nutrients but an abundant decomposer food web

95. All of the following are useful methods for reducing domestic water use EXCEPT

(A) using low-flow shower heads
(B) using low-flush volume toilets
(C) turning off water while brushing your teeth
(D) fixing leaks as soon as they start
(E) lowering the temperature of the water heater

96. Biodiversity is a direct result of which of the following?

(A) Deforestation
(B) Sanitization
(C) Respiration
(D) Erosion
(E) Evolution

97. Students studying a river found high levels of fecal coliform bacteria. They concluded that

(A) this water is fit to swim in
(B) a nearby treatment plant added chlorine to the waste water
(C) they can safely drink the water
(D) untreated animal waste was put in the water
(E) a factory was polluting the river

98. During an El Niño-Southern Oscillation, weather events change in which of the following areas?

(A) The Pacific and Indian Oceans
(B) The Atlantic and Indian Oceans
(C) The Arctic Sea
(D) The Indian and Antarctic Oceans
(E) The Atlantic and Pacific Oceans

99. Which of the following pairs correctly matches the source of gray water with its most frequent use in the home?

(A) Dishwasher and sink water used to flush toilets
(B) Flushed toilet water used to irrigate garden plants
(C) Dishwasher and sink water used to irrigate garden plants
(D) Water collected from rainfall used to flush toilets
(E) Water from showers used for dish washing

100. The Clean Water Act established all of the following guidelines EXCEPT

(A) implemented pollution control programs
(B) set water quality standards for all contaminants in surface waters
(C) made it unlawful for any person to discharge any pollutant from a point source into navigable waters
(D) demanded that an environmental impact statement be prepared for any major development
(E) funded the construction of sewage treatment plants

END OF SECTION I

GO ON TO THE NEXT PAGE.

ENVIRONMENTAL SCIENCE
Section II
Time—90 minutes
4 Questions

Directions: Answer all four questions, which are weighted equally; the suggested time is about 22 minutes for answering each question. Where calculations are required, clearly show how you arrived at your answer. Where explanation or discussion is required, support your answers with relevant information and/or specific examples.

1. A class wished to determine the LD_{50} of a particular herbicide, Chemical X. Using standard laboratory apparatus and glassware, they accurately made the following dilutions: 1.0M, 10^{-1}M, 10^{-2}M, 10^{-3}M, 10^{-4}M, and 10^{-5}M. They grew the seedlings under standard conditions, varying only the concentrations of Chemical X. Finally they determined the percentage of seedlings that germinated at each concentration.

   (a) Name a reasonable hypothesis that this experiment could test. Describe the experimental control group and give one method for performing repeated trials.

   (b) Using the axes below, graph a set of hypothetical results. Indicate the LD_{50} concentration. Properly label the axes and provide a title for the graph.

   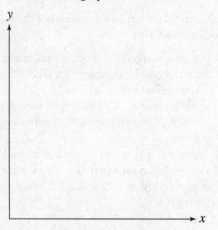

   (c) Describe one positive outcome and one negative outcome of using herbicides in the environment.

GO ON TO THE NEXT PAGE.

2. The diagram below illustrates the demographic transition model of the relationship between economic status and population.

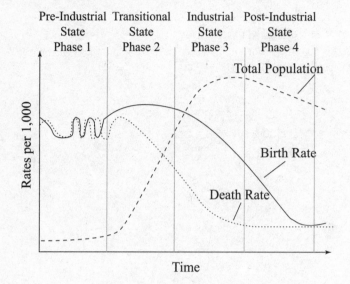

--- -- = Total Population

——— = Birth Rate

·········· = Death Rate

(a) In phases 2 and 3, there is a large difference between the birth rate and the death rate. Describe the effects on the overall population as a result of this difference. Explain why the population doubling time during these phases is short.

(b) Choose one of the four phases and describe an economic factor that would account for the differences between birth rate and death rate.

(c) Describe one biological method of birth control.

(d) Population experts have reported that in some developing countries, the population is experiencing a reverse transition from phase 2 to phase 1. Describe what would happen to a country's population and describe one event that would cause this reverse transition.

3. Under certain conditions, an internal combustion car engine produces approximately 3 grams of NO_x per kilometer driven. In Country C there are 300 million cars and each car is driven only 20,000 km per year.

(a) Calculate the number of metric tons of NO_x produced by the cars in Country C in one year. 1 metric ton = 1,000,000 g.

(b) Describe a secondary pollutant that is derived from the NO_x produced by country C, and how that pollutant travels to adjacent countries.

(c) Describe one abiotic and one biotic impact that the NO_x pollution will have on any countries adjacent to Country C.

(d) Describe one method that Country C could employ to reduce the amount of emitted NO_x.

GO ON TO THE NEXT PAGE.

4. The diagram below is of a hypothetical food web found in an estuary.

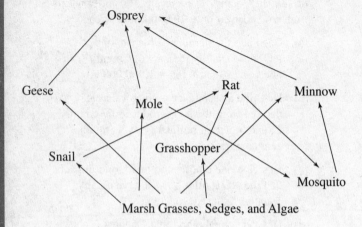

(a) Describe two abiotic factors of the aquatic component of this estuary that change twice daily.

(b) Draw an energy pyramid based on the food web. For each trophic level, give two examples of the organisms that would occupy that level.

(c) Cultural eutrophication can cause breakdowns in this type of food web. Give one example of how cultural eutrophication might occur and give one detrimental effect of this process.

STOP

END OF EXAM

_____

Chapter 17
Practice Test 2:
Answers and
Explanations

ANSWER KEY

1.	D	35.	E	69.	A
2.	B	36.	A	70.	A
3.	A	37.	D	71.	D
4.	D	38.	B	72.	A
5.	C	39.	A	73.	B
6.	D	40.	B	74.	B
7.	A	41.	B	75.	A
8.	B	42.	B	76.	C
9.	E	43.	C	77.	C
10.	C	44.	C	78.	C
11.	C	45.	D	79.	E
12.	B	46.	B	80.	C
13.	D	47.	D	81.	C
14.	E	48.	E	82.	D
15.	A	49.	E	83.	C
16.	A	50.	B	84.	C
17.	C	51.	D	85.	D
18.	B	52.	C	86.	A
19.	E	53.	E	87.	A
20.	D	54.	B	88.	B
21.	E	55.	C	89.	B
22.	B	56.	B	90.	C
23.	C	57.	D	91.	E
24.	A	58.	D	92.	A
25.	D	59.	D	93.	D
26.	B	60.	D	94.	B
27.	C	61.	B	95.	E
28.	A	62.	C	96.	E
29.	E	63.	B	97.	D
30.	D	64.	C	98.	A
31.	D	65.	E	99.	C
32.	A	66.	C	100.	D
33.	D	67.	B		
34.	A	68.	D		

EXPLANATIONS FOR SECTION I—PART A

1. **D** In predation, one organism (predator) eats another organism (prey); the relationship between a cheetah and an antelope is predator/prey.

2. **B** In mutualism, both organisms benefit from the relationship; mycorrhizae are fungi that are associated with plant roots.

3. **A** In parasitism, the host is harmed by the relationship, while the parasite benefits. It is important to note that parasites do not usually kill their hosts.

4. **D** In predation, one organism (predator) eats another organism (prey); the hawk is a carnivore that preys on the mouse, which is a herbivore.

5. **C** In competition, two organisms are competing over a resource. This usually results in the "less fit" of the two competitors losing (the principle of competitive exclusion).

6. **D** National Parks are restricted-use lands where people can only perform recreational activities such as camp, hike, fish, and boat.

7. **A** National Forests are multi-use lands where logging, grazing, farming, oil and gas extraction, mining, recreation, fishing, and hunting may all occur.

8. **B** National Resource lands are multi-use lands; among other things, they can be used to supply energy and strategic minerals. Grazing can occur as long as permits are obtained.

9. **E** National Wilderness Preservation Areas are restricted-use lands that are similar to National Parks, but with added restrictions that make the building of roads and the presence of motorized vehicles forbidden.

10. **C** National Wildlife Refuges are moderately restricted lands that are intended to provide protection for the habitats of waterfowl and big game. Hunting, trapping, and fishing can all occur on these lands with a permit.

11. **C** You should use POE on these questions—especially if you've never heard of some of these incidents! On December 2, 1984, in Bhopal, India, the Union Carbide pesticide production plant exploded, releasing 36 tons of deadly methyl isocyanate gas. 600,000 people were exposed to the pesticide, and 7,000–16,000 were killed.

12. **B** Between 1942 and 1953, Hooker Chemical dumped steel drums filled with chemical wastes into an abandoned canal in Niagara Falls, New York. In 1953, the canal was sold to the local school board, which built an elementary school; later a housing development was also built there. Chemicals leaked out of the drums, contaminating homes and the water supply. Many people, especially children, became sick, and leukemia was the predominant illness.

13. **D** During a particularly cold day in London, England, in December of 1952, the burning of coal produced a very thick gray-air smog over the city. This smog, combined with the use of diesel-powered buses, created a layer of thick smog that lasted for almost three weeks. During that time, it is estimated that there were between 2,000 and 4,000 deaths due to respiratory illnesses.

14. **E** There are five rivers that empty freshwater into the Chesapeake Bay in Maryland. These rivers all carry a heavy load of fertilizers from runoff from farms, homes, and other facilities that are in their watersheds. Unfortunately, the fertilizers cause eutrophication, which leads to very low oxygen levels. The low oxygen levels make the area unlivable for many plant and animal species.

15. **A** On April 26, 1986, engineers were running an experiment on the Chernobyl reactor in the Ukraine. This resulted in a series of explosions that exposed the core of the reactor to the outside world. An estimated 125,000 people have died as a direct result of the disaster, and about 3.5 million people have become sick in the years since the explosion.

16. **A** As rainwater passes through landfills, it dissolves compounds and carries them into the ground or along the ground, away from the landfill. This contaminated runoff is called leachate.

17. **C** In chemical treatment, an acidic compound could be treated with a basic compound in order to neutralize the substance and render it less harmful to the environment.

18. **B** During the burning of waste materials, chemical reactions may occur that yield toxic substances. These toxic substances are released into the atmosphere in the form of smoke or steam.

19. **E** The BOD (biological oxygen demand) of a body of water is a measure of bacterial activity in the water. The bacteria that live in the water subsist on organic waste; if the BOD is high, then the bacteria are using up most of the oxygen in the water, which leaves little oxygen available for use by other organisms.

20. **D** In the biological treatment of waste, bacteria and fungi are employed to decompose organic material.

21. **E** Wind turbines are placed in areas where there is a constant wind supply; but migratory birds also rely on constant winds, and may fatally encounter the turbine blades.

22. **B** In hydrogen cells, hydrogen and oxygen are catalytically combined to produce electricity, water, and some heat. Among other places, hydrogen cells are used in hybrid automobiles.

23. **C** Tidal generators are placed in areas where there is a significant tide. As the water moves in and out of a bay, for example, the force of the water spins a turbine, generating electricity.

24. **A** Photovoltaic cells use boron and silicon to convert radiant energy from the sun into an electric current. Special devices can collect the sun's rays and concentrate them on a material that either transforms the heat into electricity or uses the solar energy directly.

25. **D** Geothermal plants trap steam created by underground water that is heated by magma below the earth's crust.

26. **B** Coal contains large amounts of sulfur, which was present in the plants that formed the coal. Another contributor to the sulfur content of coal is the bacteria that decomposed the plant material long ago.

27. **C** Pollen from many different types of plants triggers an allergic response in many people.

28. **A** Ozone (O_3) is destroyed in the stratosphere in a complex set of chemical reactions involving chlorine atoms (which are found in chlorofluorocarbons, or CFCs) and the UV light from the sun.

29. **E** As spilled gasoline evaporates, the gases produced enter the atmosphere. Gasoline is composed of hydrocarbons, which are compounds made primarily of hydrogen and carbon. Evaporated fossil fuel is one component of photochemical smog.

30. **D** Particles of soot can range in size from 0.001 μm to 100.0 μm. They can remain airborne and enter the lungs during breathing; it can also enter through the eyes. Soot can cause eye redness and even asthma.

EXPLANATIONS FOR SECTION I—PART B

31. **D** Noise is measured in decibels (dB). A vacuum cleaner has a loudness of about 70 dB, a chain saw has a loudness of about 100 dB, the average factory has a loudness of about 80 dB, a rifle has a loudness of about 160 dB, and a lawn mower has a loudness of about 90 dB. The rifle would most likely damage the hearing because it is the loudest.

32. **A** Population density is a measure of individuals per unit area. On land, it is measured as the number of individuals per square area (hectare, meter, etc.), while in aquatic environments, the measure is based on the number of individuals per unit of volume (milliliters, liters, etc.).

33. **D** This graph illustrates what can happen to supplies of a nonrenewable resource, such as oil. Curve A shows a resource which is being rapidly consumed, which is not being recycled, and for which no new reserves are found. Curve B shows a resource which can be partially recycled, for which there are more reserves, and which is being conserved. Curve C is seen when the use of the resource is reduced and there are relatively ample reserves.

34. **A** The Aral Sea, located in Uzbekistan and Kazakhstan (both countries were part of the former Soviet Union), is a saline lake. It is in the center of a large, flat desert basin. In the past few decades, the Aral Sea's volume has decreased by 75 percent of both natural and human causes. (B) would be a result, not a cause; (C) is not correct because earthquakes do not cause water loss; (D) and (E) are not correct because pesticides would cause water pollution, not water loss.

35. **E** All three of these events can occur when fish populations are over-harvested. The U.S. National Fish and Wildlife Foundation found that 14 major commercial fish species are almost depleted. Also, one-fourth of the annual fish catch is by-catch or nontarget organisms.

36. **A** In flood irrigation, water flows via gravity to all parts of the field. In this process, as little as 60 percent of the water is available to plants because of processes such as evaporation, runoff, and seepage. Drip irrigation uses above- or below-ground pipes to deliver water directly to the plant's roots; this method is highly effective. Low-pressure spray heads deliver large drops of water (which evaporate slowly) to plants. This process is more than 90 percent efficient.

37. **D** Watch your letters! Option B is (D). (D) represents the exponential growth phase of this population. In exponential growth, birth rates are high and environmental resistance is low, so the population rapidly increases over a short period of time. Also note that the line is at its steepest slope.

38. **B.** In population graphs such as this logistical growth curve, the label K represents the carrying capacity. The carrying capacity describes the maximum number of individuals that the habitat can sustain for a long time. The biotic potential (or how rapidly the organisms can reproduce) is balanced by the environmental

resistance (factors that lower a population). Note how the population size modulates above and below the carrying capacity.

39. **A** Population growth is studied over time, so the independent axis (x) is time. (B) is the dependent variable, population number, which is on the vertical axis (y). (C) is the name of the growth pattern; it is not a variable. (D) refers to environmental effects; factors that would slow down population growth. (E) refers to the number of babies born per 1,000 people.

40. **B** "Environmental resistance" refers to those factors that slow the growth of a population. Unfavorable abiotic factors (e.g., the climate is too hot or too dry) or biotic factors (e.g., predators or disease) will limit the number of survivors. At point D, the population is about to increase, so environmental resistance is not high—it's low.

41. **B** Phosphates are most commonly found as phosphate salts containing phosphate ions (PO_4^{-3}). Phosphorous does not dissolve easily in water and doesn't form a gas at normal temperatures and pressures. Most of the earth's sulfur is stored in rocks as sulfate salts (SO_4^{-2}). Both phosphorus and sulfur are only held in living organisms for short periods of time.

42. **B** The movement of tectonic plates in the earth's lithosphere occurs because these plates are floating on the semi-liquid magma underneath them. (A) and (C) both deal with the loss of species, and (D) and (E) deal with the movement of populations.

43. **C** The Green Revolution started in the 1950s; farmers used new methods of farming, such as breeding high-yield crops, better irrigation processes, and large amounts of fertilizer and pesticides.

44. **C** Mortality figures are calculated as the number of deaths per 1,000 people aged zero to one. For example, in 1987, 9,889 babies were born in state X. In that same year, 116 infants (aged zero to one) died in the state. The infant mortality rate would be calculated as

$(116 \div 9,889 = .0117) \times 1,000 = 11.7$

So, the mortality rate is 11.7.

45. **D** Underground storage tanks would leak chemicals into the surrounding ground; these chemicals might then move into the water that lies underground, also known as the aquifer. (A) may pollute aquifers as well, so it is not a good answer. (B) is not associated with underground events. (C) is the boundary between two habitats; it is too broad an answer. (E) is too broad a term; it includes the liquid, solid, and gaseous forms of water.

46. **B** Let's go through all of the answer choices one by one. (A): CITES is an agreement that concerns the international trade of endangered species. (B): The Kyoto Protocol is the correct answer. (C): The Montreal Protocol concerns the emissions of CFCs. (D): The Clean Air Act deals with air pollution in the United States; it is not an international treaty. (E): The RCRA is an act that concerns the management of mineral resources.

47. **D** (D) is the best answer. (A), (C), and (E) all require the use of more mineral (and therefore nonrenewable) resources. (B) consumes mineral resources and is also costly. Sustainability is the ability of people to utilize a resource over the long term (for many generations). This can most economically be achieved by reducing demand and finding ways to reduce waste as much as possible. Replacing incandescent lightbulbs with florescent lightbulbs is one example of a way to decrease waste.

48. **E** The "Energy Policy Conservation Act," enacted into law by Congress in 1975, added Title V, "Improving Automotive Efficiency," to the Motor Vehicle Information and Cost Savings Act and established CAFE standards for passenger cars and light trucks. The Act was passed in response to the 1973–1974 Arab oil embargo. The near-term goal was to double new car fuel economy by model year 1985. (E) clearly states the goal of the Corporate Average Fuel Economy policy.

49. **E** DDT is an organic insecticide first prepared in 1873. It is a neurotoxin and causes nerve cells to fire continuously, leading to spasms and death. It is dangerous because it can accumulate in the fatty tissues of all animals, and humans that are exposed to it can become sick. Birds exposed to DDT lay eggs with shells that are too thin for the embryo to survive incubation.

50. **B** This is a question about graph interpretation. In 1950, the developing countries had a population of approximately 2 billion people. In 2050, the population is expected to be about 6 billion people. The difference of 4 billion is the correct answer.

51. **D** This graph interpretation question asks you to observe what contributes most to the rise in world populations. The developed nations' overall population rises very slowly. On the other hand, in developing countries, there is a rapid rise in the number of people. This would contribute to most of the total population.

52. **C** In the year 2000, the less developed countries had a population of approximately 4 billion people. During that same year, the developed countries collectively contained approximately 1 billion people. The best answer is (C).

53. **E** Domestic water use includes all uses of water around the home, and the chart indicates that it is responsible for 1 percent of all water use.

54. **B** Sea levels are rising because water is entering the oceans from melting ice caps and glacial melting; another contributor to rising sea levels is the thermal expansion of water. (A) would occur if there were less ozone in the atmosphere. (C) is not correct: Radon seepage does not correlate to atmospheric temperatures. (D) would occur if there were lower temperatures, not higher temperatures. (E) is incorrect; it would be a result of global cooling.

55. **C** Ozone is the one gas that does not absorb infrared radiation that's reflected from Earth. Infrared radiation is the energy that heats the atmosphere and leads to global warming. Water vapor is the most common of the gases listed and it does absorb infrared (IR) radiation. It is not considered a "global warming" gas because it's so common and because its atmospheric levels have remained constant for the last 160,000 years.

56. **B** Poverty is a force that drives many people to exploit the land they live on. Farming, trade in animals, and building roads are all methods that people use to make money.

57. **D** The term "per capita" translates from the Latin to mean "per head" or "per person." This concept is important in areas such as resource consumption, standard of living, and other population studies.

58. **D** When two of these massive rock plates collide, often one slips under the other; this process is called subduction. The collision of two plates causes the rock layer of the lithosphere to crack. At these sites, magma may rise from the molten exterior of the earth, and a volcano is formed. (A) is a human activity; (B) can be caused by many geological processes; (C) is an effect of the relationship between the earth and the sun; and (E) is a biological process.

59. **D** "r" and "K" describe the different ways that populations reproduce. r-strategists reproduce at their biological potential, the maximum reproductive rate. These populations produce as many offspring as possible, and many of the offspring will not live to reproductive age. These populations are characterized by rapid increases and rapid declines, for example, as seen in insect populations and weeds. K-strategists produce few offspring and nurture them carefully; humans are examples of K-strategists.

60. **D** Convectional cooling is an important process, moving heat and moisture around the planet. As warm, moist air rises, it cools. This cooling causes the water vapor to condense and fall as rain. The now cool, dry air becomes denser as it sinks to the earth's surface, where more moisture and heat are picked up and convection starts again.

61. **B** Sedimentary rock is formed by the compression of eroded rock, silt, and the remains of dead organisms. Under heat, pressure, and stress within the mantle, sedimentary rock can form metamorphic rock. Deeper in the mantle, the metamorphic rock melts and cools, forming igneous rock.

62. **C** Since sedimentary rock is made of the remains of dead organisms, it would contain the greatest number of fossils. (E) is not correct because the melting and cooling process that forms igneous rock would destroy fossils, which are usually quite fragile.

63. **B** The North Atlantic current circulates water that was warmed by the Sun (at the equator) into the northern latitudes, thus warming the land masses. When it cools, the water becomes more dense and it picks up large amounts of CO_2 and salt. This cold, salty water circulates into the equatorial areas, cooling them while releasing CO_2 and salt.

64. **C** (A) and (B) are gases formed by combustion. (D) is a gas produced by the radioactive decay of uranium. (E) is a gas that comes from furniture stuffing and foam insulation. Asbestos is a mineral used in insulating pipes and as a fire retardant. If not completely confined, it can crumble and the fibers can be inhaled into the lungs, which can cause certain types of cancer.

65. **E** Chlorine is carried into the stratosphere via chlorofluorocarbons (CFCs). These compounds release Cl as a result of reactions that occur in the stratosphere. Each Cl atom can catalyze the destruction of millions of ozone molecules. (A): Fluorine is a component of CFCs but does not enter the reaction with the ozone. (B): Oxygen is a component of ozone, not a catalyst of its destruction. (C): Mercury is toxic to animals and plants. (D): Platinum is a component in catalytic converters.

66. **C** Transpiration is the loss of water to air, from plants. Transpiration adds a lot of moisture to the air; this can influence local environmental conditions, such as rainfall and temperature. Transpiration occurs as water molecules evaporate through the openings in plant leaves.

67. **B** Let's go through the answer choices. (A) is used to turn steam into water so that it can be turned back into steam and spin the turbines. (C) is the site where the heat from the core heats water, turning it into steam that spins the turbines. (D) contain uranium, which is the fuel. Turbines, (E), generate electrical power and are not directly connected to the reactor. Control rods contain compounds, such as cadmium, that absorb neutrons; this reduces their ability to perpetuate a chain reaction.

68. **D** Riparian zones are the areas of vegetation that abut a river or stream. They form a corridor of vegetation along the banks of the river or stream. They increase habitat diversity for organisms living in or near the water and can absorb excessive nitrogen, phosphorous, and pesticides, preventing them from entering the water.

69. **A** Trawling involves dragging a net across the ocean floor. This disrupts the habitat and catches a wide variety of bottom-dwelling species. Drift nets float in the water; long lines include baited hooks, and they are towed behind boats; purse-seine catches surface fish as a net is drawn up from below; and fish farming is the growth of fish, from egg-stage to the point of harvest, in confined pens.

70. **A** In 1987, 36 nations met in Montreal and signed the Montreal Protocol, which cut the emissions of CFCs by about 35 percent between 1989 and 2000. In 1992, the protocol was updated in a meeting in Denmark to accelerate the phasing out of ozone-depleting chemicals. (B): Kyoto Protocol dealt with CO_2 levels. (C): CAA affected the United States only. (D): The Rio Earth summit dealt with many global environmental problems, but Montreal was specific to ozone depletion. (E): This refers to the U.S. "Superfund" Act.

71. **D** Bioaccumulation is the process by which chemicals remain and accumulate in the bodies of animals. These chemicals come from food and cannot be metabolically removed from the tissues. Generally, they are fat-soluble in nature. Biomagnification occurs when a predator eats several organisms and each individual prey has a bit of the chemical in its tissues. Bioaccumulation and biomagnification work together to increase the levels of toxic materials in the bodies of animals. Bioremediation is the removal of toxic compounds by living organisms.

72. **A** A reduction in the creation of pollution is always the least expensive method. (E) is impractical because air pollution can still spread to other areas. (D) does not help acid deposition on land and in the atmosphere. (C) is not possible, and (B), taller stacks would just spread pollution to other locations.

73. **B** Gamma radiation can penetrate most materials, and of the listed forms of radiation, is most harmful to humans. When it gets into a cell and damages the DNA, cancer can result. (A): Alpha rays can be stopped by a piece of paper. (C): Beta rays can be stopped by wood or clothing. (D): Infrared radiation is heat, as comes from a stove. (E): Radon is not a form of radiation; it is an element that can produce radiation.

74. **B** The producers in the deep sea hydrothermal vent ecosystem do not capture sunlight and use it to perform photosynthesis. Instead, they use chemical energy from the hot mineral-rich water that comes out of the vents. The producers in these vents are bacteria that fall into the domain Archaea. The other animals that live near the vents; tube worms, crabs, and many others, all depend on the bacteria for food. (A), (C), (D), and (E) all refer to ecosystems in which producers use radiant energy to carry out photosynthesis.

75. **A** All of the answer choices represent correct processes except (A): adding it to beverages. In this process, CO_2 would reenter the atmosphere, which is opposite to the goal of sequestering, which is long-term storage.

76. **C** (C) is not necessarily true. Invasive species are r-selected organisms; they are usually small adults, reproduce many offspring, sexually mature very quickly, and are generalists. (A), (B), (D), and (E) all correctly describe invasive species.

77. **C** (A) is a description of the First Law of Thermodynamics. (B) describes the law of mass conservation. (D): The usefulness of electricity is the same regardless of how it is produced. (E) is incorrect; it is a violation of the First Law of Thermodynamics. The Second Law of Thermodynamics states that, at each energy transformation, there is a loss of energy.

78. **C** When large amounts of river water are used to irrigate fields, the percolation and evaporation of the water leaves behind the salts. The dissolved salts accumulate over many years and make the soil less amenable to plant growth.

79. **E** Many of the young die in species that are considered early-loss; one example of this is seen in fish. In some fish species, most of the individuals die in the first weeks after hatching. (A) describes species such as birds, and (B), (C), and (D) all describe late-loss species, like humans and elephants.

80. **C** Secondary succession is defined as succession that begins on land that was disturbed by an (often human) activity. Primary succession begins on land that was exposed to abiotic factors; for example, when lava cools or when a glacier leaves an area.

81. **C** By definition, early stage species are small, and exhibit little diversity. At the early stages of succession, an area is populated mostly by r-selected organisms. Mid-stage and late-stage communities generally have larger adult species and many more k-selected organisms.

82. **D** This practice is called a "debt for nature swap." In it, participating countries act as guardians for forests or other biomes in return for foreign aid or debt relief. It has been successfully done in Guyana.

83. **C** Steady winds create polar vortices, which can trap large amounts of CFCs from other parts of the world. In winter, ice crystals form in the upper atmosphere. The CFCs accumulate on the crystals and when the polar spring returns, the crystals melt and release the CFCs, then the free Cl breaks down the ozone.

84. **C** The crown is the top, or most rapidly growing part, of the tree. In very large forest fires the crowns of the trees burn, weakening or killing the trees. When leaves or needles burn in smaller fires, this takes away fuel for future fires and breaks down matter to ash, which provides nutrients for the trees. Fires are also responsible for allowing the seeds of many plant species to germinate.

85. **D** Dams can have detrimental impacts on local environments and populations. These detrimental effects include: high construction costs, high CO_2 emissions from decaying biomass, the flooding of natural areas, the conversion of terrestrial ecosystems to aquatic ecosystems, danger of collapse, and blocking migratory fish. Choice II, the generation of low pollution electricity, is a positive effect, so eliminate any answer choices that include II.

86. **A** Evolution is the change in the genetic makeup of a population over time. The key term is genetic makeup—also referred to as the gene pool. In this type of question, look for a phrase or word that implies the passage of time; in this question, it's "becomes." That should give you a hint that at one time the population was not resistant and now it is resistant. All the other options (except D) are temporary changes. (D) is incorrect because the change occurs in one individual, not a whole population.

87. **A** The change in form of a substance, from solid, to liquid, to gas, or back again, is referred to as a phase change. Phase changes do not change the number or types of atoms that compose the molecule.

88. **B** Ozone formation occurs when sunlight (UV) binds an atom of oxygen to a diatomic oxygen molecule (O_2). Neither radioactive decay, nor magma's heat, influences events in the stratosphere. Wind energy does not cause the chemical reaction between O_2 and oxygen.

89. **B** Ozone is a major component of air pollution, especially pollution that results from the combustion of fossil fuels. In the stratosphere, however, ozone blocks large amounts of UV light from the sun. (A): All animals breathe O_2, so it is vital to survival. (C): Chlorofluorocarbons destroy O_3, so it is harmful in the stratosphere. (D): H_2SO_4, or sulfuric acid, is harmful anywhere it's found in the atmosphere. (E): DDT is an insecticide. It can cause lung irritation, but is not found in the stratosphere.

90. **C** Primary pollutants are created via combustion (coal burning, wood burning, etc.) or from volcanic activity. Examples of pollutants produced this way include oxides of sulfur and nitrogen. Secondary pollutants are created when primary pollutants combine in the atmosphere. These reactions are very complex, and involve energy from the sun, water vapor, and the mixing effects of air currents.

91. **E** The calculation for power consumption is

$$\text{watt-hours} = \text{volts} \times \text{amps} \times \text{time}$$
$$\text{watt-hours} = 120 \times 10 \times 1 = 1,200$$

92. **A** (B), (C), and (E) are all geological processes. For (D), radioactive materials do not form fossil fuels. Coal and other fossil fuels all formed from plants and other organic material that lived some 300 to 400 million years ago. These plants and animals decayed and the remains were exposed to tremendous pressures and temperatures.

93. **D** Countries that have more food will have healthier children and adult populations, and thus they will have lower death rates. Increased food production results in increased erosion, because after crop harvests, the bare ground is more readily eroded. Biodiversity also suffers, as complex native communities are turned over to monoculture (single species) crop fields. Approximately 90 percent of applied pesticides do not reach their target, and can end up in water and soil. The heavy use of large machinery means that more plants can be harvested, but these machines are powered by fossil fuels.

94. **B** Mid-latitude grasslands alternate between times of rapid growth and drought. Drought and the cold winter temperatures permit only a small amount of decomposition each year, which leads to the accumulation of organic matter. In North America, this layer can be more than 30m thick. (A) describes coniferous forest soils, (C) is arctic tundra, (D) describes desert soil, and (E) refers to tropical rain forest soil.

95. **E** (E) might reduce the cost of heating the water, but it will not slow down water consumption. All of the other answer choices describe ways of lowering the use of water.

96. **E** Evolution results in the development of populations that are fit to live in certain habitats. Sexual reproduction allows for genetic diversity in populations. Organisms with slightly different adaptations can survive in different niches within a habitat. So, as more different species populations evolve in different niches, the biodiversity in a habitat increases.

97. **D** Because fecal coliform bacteria come from animal (and human) waste, the answer is (D). (A) and (C) are not correct, as fecal bacteria can cause a number of diseases. (E) is too general to be the best answer. (B): Chlorine is used to kill fecal bacteria.

98. **A** In an El Niño, the winds that normally blow from the eastern Pacific to the western Pacific weaken or stop. This shifts the precipitation from the western Pacific to the central Pacific and can alter weather events over as much as two-thirds of the globe.

99. **C** Gray water is water from sources such as washing machines, showers, dishwashers—almost any water-using appliances, except toilets. In areas of limited rainfall, gray water is stored in underground tanks and used as irrigation water for gardens, lawns, etc. It may constitute 50 percent to 80 percent of domestic wastewater. Generally, it is not used to irrigate crop plants.

100. **D** (A), (B), (C), and (E) are all provisions of the CWA. (D) is a provision of the National Environmental Policy Act, so it is the exception and the correct answer.

FREE-RESPONSE EXPLANATIONS

Question 1

A class wished to determine the LD_{50} of a particular herbicide, Chemical X. Using standard laboratory apparatus and glassware, they accurately made the following dilutions: 1.0M, 10–1M, 10–2M, 10–3M, 10–4M, and 10–5M. They grew the seedlings under standard conditions, varying only the concentrations of Chemical X. Finally they determined the percentage of seedlings that germinated at each concentration.

(a) Name a reasonable hypothesis that this experiment could test. Describe the experimental control group and give one method for performing repeated trials.

You should have provided a clearly written statement that describes how the independent variable (Chemical X concentration) will affect the dependent variable (germination of seeds). Some examples might include: "As the concentration of Chemical X increases, the amount (or percentage) of seed germination decreases"; "As the amount of Chemical X decreases, the amount of seed germination increases"; or "The concentration of Chemical X does not affect the rate of seed germination."

The control group would include seeds grown under exactly the same conditions as the experimental group, except that they would have no Chemical X added to their environment.

You should indicate that the seeds were germinated under the various conditions. Trials might include many seeds in one container, or many containers each with a single seed, all watered with the same amount of Chemical X.
(3 points maximum—1 point each for hypothesis, control, and repeated trials)

(b) Using the axes below, graph a set of hypothetical results. Indicate the LD_{50} concentration. Properly label the axes and provide a title for the graph.

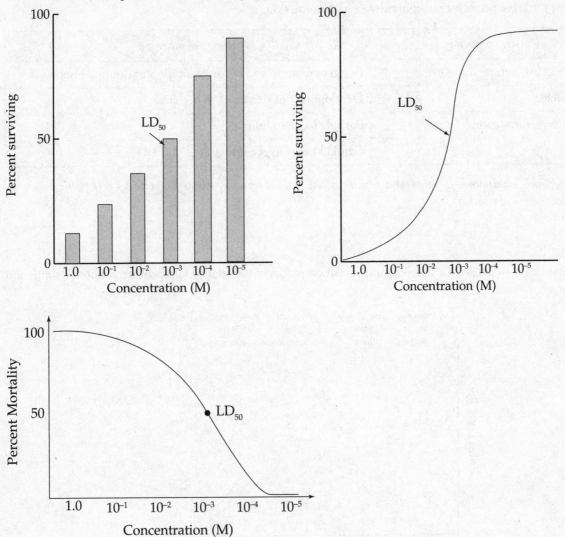

The independent variable (*x*-axis) is concentration (1.0M, 10–1M, 10–2M, 10–3M, 10–4M, and 10–5M). The dependent variable (*y*-axis) is percent surviving. The LD_{50} value is indicated by the point at which less than 50 percent of the seeds germinated. Students can use either a histogram (left) or line graph (right) for correct graph type.

(4 points maximum—1 point for correct x-axis, 1 point for correct y-axis, 1 point for reasonable line or histogram graph data, and 1 point for illustrating LD_{50})

(c) Describe one positive outcome and one negative outcome of using herbicides in the environment.

See the table below for possible answers to this question.

Positive outcome	Negative outcome
Lower costs for crop production	Bioaccumulation and biomagnification of deadly poisons
Higher yields	Development of resistance to herbicide
Rapidly treats infestations	Possible harm to humans
	Can kill non-target organisms

(3 points maximum—1 point for a positive effect, 1 point for a negative effect, and up to 1 point for a good description of either)

Question 2

The diagram below illustrates the demographic transition model of the relationship between economic status and population.

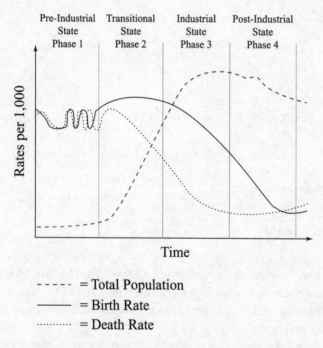

(a) In phases 2 and 3, there is a large difference between the birth rate and the death rate. Describe the effects on the overall population as a result of this difference. Explain why the population doubling time during these phases is short.

With a large number of children being born and fewer dying, the population would grow very rapidly. Doubling times would be very short because the percent growth would be very large. The Rule of 70 states that when the percent change is large, the doubling time is rapid (e.g., 70/4 percent = 17.5 years doubling time, while 70/1 percent = 70 years doubling time). When the large number of children mature and procreate, there will be even more children born. This is known as population inertia. *(2 points maximum—1 for effects and 1 for doubling time effect)*

(b) Choose one of the four phases and describe an economic factor that would account for the differences between birth rate and death rate.

- Phase 1—Birth rates (BR) and death rates (DR) are approximately equal. While there are a large number of births, they are offset by high death rates due to starvation, disease, unsanitary water, poverty, or any factor that would increase the death rate.

- Phase 2—The death rate falls due to better economic conditions (more food, better health care, more jobs means there is more money to spend). The birth rate is still high.

- Phase 3—The birth rate begins to drop due to more education opportunities for women, an increase in the cost of a child's education, an increase in the cost of raising children, more available and effective forms of birth control, fewer children needed for child labor, more children surviving past infancy, and any other factor that would lower birth rate.

- Phase 4—The birth rate and death rate are now about equal. Any factor that lowers the birth rate and maintains a low death rate is acceptable.

  (2 points maximum—1 point for description and 1 point for explanation)

(c) Describe one biological method of birth control.

There are three basic methods.

- Surgical—In females, examples include tubal ligation or cutting of oviducts (Fallopian tubes), hysterectomy (removal of the uterus), and removal of the ovaries. In males, examples include removal of the testicles and vasectomy (cutting of the vas deferens).

- Hormonal—Birth control hormones for women include "the pill" and variations, such as hormonal implants. In men, hormonal methods and chemical control are still under development.

- Obstructive/Non-hormonal Chemical—Methods include cervical cap, condoms, abstinence, vaginal ring, intrauterine device, vaginal pouch, diaphragm, spermicide, rhythm method, withdrawal, and douche.
  (2 points maximum—1 for method and 1 for explanation)

(d) Population experts have reported that in some developing countries the population is experiencing a reverse transition from phase 2 to phase 1. Describe what would happen to a country's population as a result of such a transition, and describe one event that would cause this reverse transition.

A backward transition would mean that the death rate is increasing while a high birth rate is maintained. This means that the overall population would grow very slowly, or even decrease, depending on how close the birth rate and death rate are to each other. Events that might cause a rapid rise in death rate are war, civil unrest, abject poverty, a rapid rise in infectious diseases (such as AIDS), or natural disaster.
(4 points maximum—2 for description of effect and 2 for naming the event and explanation)

Question 3

Under certain conditions, an internal combustion car engine produces approximately 3 grams of NO_x per kilometer driven. In Country C there are 300 million cars and each car is driven only 20,000 km per year.

(a) Calculate the number of metric tons of NO_x produced by the cars in Country C in one year. 1 metric ton = 1,000,000 g.

3×10^8 cars $\times 2 \times 10^4$ km per year $\times 3$ g NO_x per km $= 18 \times 10^{12}$ or 1.8×10^{13} g NO_x. 1 metric ton is 1×10^6 g, so the final answer is 1.8×10^7 tons of NO_x.
(2 points maximum—1 point for setup and 1 point for correct answer)

(b) Describe a secondary pollutant that is derived from the NO_x produced by Country C, and how that pollutant travels to adjacent countries.

Nitric acid is a secondary pollutant produced by the addition of NO_x to atmospheric water. This acidified water is carried by wind and rain to the countries downwind and released as acid deposition (in the form of acid rain, snow, fog, sleet, or hail.) PANS (peroxyacetyl nitrates) result from NO reacting with hydrocarbons.
(2 points maximum—1 point for description of nitric acid production and 1 point for description of acid deposition)

(c) Describe one abiotic and one biotic impact that the NO_x pollution will have on any countries adjacent to Country C.

See the table below for possible answers to this question.

Abiotic effects	Biotic effects
Water Acidification	Death of young fish; loss of primary productivity in bodies of water
Soil Acidification	Death of decomposers in soil (bacteria, fungi)
Rain Acidification: Acid may burn plant leaves or bark, making plants more susceptible to disease	Death of mycorrhize living with plant roots and loss of nutrients to plants
Increased transpiration might lead to lower soil water levels	Loss of cutin layer on leaves means more transpiration out of plants

(4 points maximum—2 points for abiotic factor and description and 1 point each for biotic factor and description)

(d) Describe one method that Country C could employ to reduce the amount of emitted NO_x.

Some methods Country C could use include

- Using cleaner-burning fuels

- Adding or increasing efficiency of catalytic converters

- Increasing public transportation in order to lower the number of cars on roadways

- Taxing or employing other incentives to get people to use alternate-fuel vehicles

(2 points maximum—1 point for method and 1 point for description)

Question 4

The diagram below is of a hypothetical food web found in an estuary.

(a) Describe two abiotic factors of the aquatic component of this estuary that change twice daily.

You could have mentioned the following factors:

- Tide—High tide brings in a large volume of water. This submerges some plant life and washes in organic and inorganic nutrients. Low tide exposes some plants to air and light, so they can undergo photosynthesis; it can also wash out materials and pollutants. Some living organisms (such as jellyfish) also move with the tide.

- The dissolved O_2 (DO) content of water—The DO would rise as more water and sunlight stimulated photosynthesis in green plants and algae. It can also fall after sunset because there is no light for photosynthesis. The amount of nutrients also affects DO. With a lot of nutrients, the biological oxygen demand might rise, and DO levels go down.

- Water temperature—Temperatures will rise and fall with the rising and setting of the sun.

- Salinity—This will rise and fall with the tides. As the tide rises, more salt-bearing ocean water enters the estuary. As the tide recedes, more fresh river water enters the estuary, which lowers the salinity.

(2 points maximum—1 for each factor and its description)

(b) Draw an energy pyramid based on the food web. For each trophic level, give two examples of the organisms that would occupy that level.

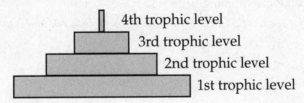

4th trophic level
3rd trophic level
2nd trophic level
1st trophic level

1st trophic level: marsh grasses, sedges, and algae

2nd trophic level: snails, geese, moles, grasshoppers, minnows, mosquitoes

3rd trophic level: rats, mosquitoes, minnows, moles, osprey

4th trophic level: osprey, mosquitoes

Note that some organisms can be part of two different trophic levels at the same time, depending on the food they're eating.

(4 points maximum—2 for the levels and 2 for the correct examples)

(c) Cultural eutrophication can cause breakdowns in this type of food web. Give one example of how cultural eutrophication might occur, and give one detrimental effect of the process.

Cultural eutrophication is the process in which human-related activities add excessive nutrients (especially nitrogen and phosphorous) to bodies of water. These excessive nutrients stimulate algal bloom. When the algae die, bacteria living in the water decompose them. These bacteria consume the water's dissolved oxygen, so that dissolved oxygen is not available to fish and other organisms. This causes the death of many aquatic organisms.

Cultural eutrophication can result from the following:

- Fertilizer runoff (farm and lawn)

- Factory discharge of nitrogenous waste

- Discharge of raw sewage

- Runoff from parking lots or other paved areas

- Dumping of waste or garbage into the water

The effects of cultural eutrophication can include

- Lower primary production in the water because the cloudy water does not let light pass to bottom-dwelling plants.

- Low levels of dissolved oxygen, which kills off aquatic organisms like fish and amphibians.

- Water that's unfit for drinking, swimming, or fishing.
  (4 points maximum—2 for the description of the process of eutrophication and 2 for an effect and its description)

HOW TO SCORE PRACTICE TEST 2

Section I: Multiple-Choice

_____ × .9000 = _____
Number Correct   Weighted
(out of 100)   Section I Score
                                  (Do not round)

Section II: Free Response

Question 1   _____ × 1.5000 = _____
                  (out of 10)   (Do not round)

Question 2   _____ × 1.5000 = _____
                  (out of 10)   (Do not round)

Question 3   _____ × 1.5000 = _____
                  (out of 10)   (Do not round)

Question 4   _____ × 1.5000 = _____
                  (out of 10)   (Do not round)

AP Score Conversion Chart
Environmental Science

Composite Score Range	AP Score
105–150	5
86–104	4
72–83	3
59–71	2
0–58	1

Sum = _____
            Weighted
            Section II Score
            (Do not round)

Composite Score

_____ + _____ + _____
Weighted   Weighted   Composite Score
Section I Score   Section II Score   (Round to nearest
                                            whole number)

ABOUT THE AUTHORS

Dr. Angela Baker is an Assistant Professor of Research at the University of Northern Colorado, working with the Biology/Environmental Studies Programs and the Math and Science Learning Center. She taught AP Biology for three years at an academic magnet school and has been involved with AP Environmental Science since its beginnings. Dr. Baker is a reader and table leader for the AP Environmental Science Exam as well as the author of several laboratories that are available on AP Central. She has directed a number of teacher workshops for the AP Environmental Science Exam. Her article *Public Health and Urban Ecology* was featured on *AP Central* (November 2003).

As an ultrastructural biologist (Ph.D. Botany/Microbiology), her research interest is the reduction of pollutants in wood pulping using nanotechnology. She and her husband live in Boulder, Colorado, and take advantage of the city's green spaces while enjoying horseback riding, cycling, camping, and canoeing.

Tim Ligget lives in Chester Springs, PA. He has taught Environmental Science at the AP and college-bound level for the last nine years at Conestoga High School. He has been a table leader and reader for the Environmental Science Exam, as well as a reader for the AP Biology exam. Prior to teaching at Conestoga High School, he worked for Educational Testing Service on several nationally administered examinations. He was a member of the test development staff during the development of the AP Environmental Science program. His interests include science fiction literature, hiking, kayaking, and sailing.

Diagnostic Test Form ○ Side 1

1. YOUR NAME: _____
(Print)   Last   First   M.I.

SIGNATURE: _____ **DATE:** ___/___/___

HOME ADDRESS: _____
(Print)   Number and Street

E-MAIL: _____
City   State   Zip

PHONE NO.: _____ **SCHOOL:** _____ **CLASS OF:** _____
(Print)

IMPORTANT: Please fill in these boxes exactly as shown on the back cover of your test book. ▼ ▼

2. TEST FORM
_____

5. YOUR NAME

First 4 letters of last name				FIRST INIT	MID INIT
Ⓐ	Ⓐ	Ⓐ	Ⓐ	Ⓐ	Ⓐ
Ⓑ	Ⓑ	Ⓑ	Ⓑ	Ⓑ	Ⓑ
Ⓒ	Ⓒ	Ⓒ	Ⓒ	Ⓒ	Ⓒ
Ⓓ	Ⓓ	Ⓓ	Ⓓ	Ⓓ	Ⓓ
Ⓔ	Ⓔ	Ⓔ	Ⓔ	Ⓔ	Ⓔ
Ⓕ	Ⓕ	Ⓕ	Ⓕ	Ⓕ	Ⓕ
Ⓖ	Ⓖ	Ⓖ	Ⓖ	Ⓖ	Ⓖ
Ⓗ	Ⓗ	Ⓗ	Ⓗ	Ⓗ	Ⓗ
Ⓘ	Ⓘ	Ⓘ	Ⓘ	Ⓘ	Ⓘ
Ⓙ	Ⓙ	Ⓙ	Ⓙ	Ⓙ	Ⓙ
Ⓚ	Ⓚ	Ⓚ	Ⓚ	Ⓚ	Ⓚ
Ⓛ	Ⓛ	Ⓛ	Ⓛ	Ⓛ	Ⓛ
Ⓜ	Ⓜ	Ⓜ	Ⓜ	Ⓜ	Ⓜ
Ⓝ	Ⓝ	Ⓝ	Ⓝ	Ⓝ	Ⓝ
Ⓞ	Ⓞ	Ⓞ	Ⓞ	Ⓞ	Ⓞ
Ⓟ	Ⓟ	Ⓟ	Ⓟ	Ⓟ	Ⓟ
Ⓠ	Ⓠ	Ⓠ	Ⓠ	Ⓠ	Ⓠ
Ⓡ	Ⓡ	Ⓡ	Ⓡ	Ⓡ	Ⓡ
Ⓢ	Ⓢ	Ⓢ	Ⓢ	Ⓢ	Ⓢ
Ⓣ	Ⓣ	Ⓣ	Ⓣ	Ⓣ	Ⓣ
Ⓤ	Ⓤ	Ⓤ	Ⓤ	Ⓤ	Ⓤ
Ⓥ	Ⓥ	Ⓥ	Ⓥ	Ⓥ	Ⓥ
Ⓦ	Ⓦ	Ⓦ	Ⓦ	Ⓦ	Ⓦ
Ⓧ	Ⓧ	Ⓧ	Ⓧ	Ⓧ	Ⓧ
Ⓨ	Ⓨ	Ⓨ	Ⓨ	Ⓨ	Ⓨ
Ⓩ	Ⓩ	Ⓩ	Ⓩ	Ⓩ	Ⓩ

3. TEST CODE   4. PHONE NUMBER

Digits 0–9 bubble columns.

6. DATE OF BIRTH

MONTH		DAY		YEAR	
○ JAN					
○ FEB					
○ MAR	⓪	⓪	⓪	⓪	
○ APR	①	①	①	①	
○ MAY	②	②	②	②	
○ JUN	③	③	③	③	
○ JUL		④	④	④	
○ AUG		⑤	⑤	⑤	
○ SEP		⑥	⑥	⑥	
○ OCT		⑦	⑦	⑦	
○ NOV		⑧	⑧	⑧	
○ DEC		⑨	⑨	⑨	

7. SEX
○ MALE
○ FEMALE

8. OTHER
1 Ⓐ Ⓑ Ⓒ Ⓓ Ⓔ
2 Ⓐ Ⓑ Ⓒ Ⓓ Ⓔ
3 Ⓐ Ⓑ Ⓒ Ⓓ Ⓔ

Begin with number 1 for each new section of the test. Leave blank any extra answer spaces.

SECTION 1

1 Ⓐ Ⓑ Ⓒ Ⓓ Ⓔ	26 Ⓐ Ⓑ Ⓒ Ⓓ Ⓔ	51 Ⓐ Ⓑ Ⓒ Ⓓ Ⓔ	76 Ⓐ Ⓑ Ⓒ Ⓓ Ⓔ
2 Ⓐ Ⓑ Ⓒ Ⓓ Ⓔ	27 Ⓐ Ⓑ Ⓒ Ⓓ Ⓔ	52 Ⓐ Ⓑ Ⓒ Ⓓ Ⓔ	77 Ⓐ Ⓑ Ⓒ Ⓓ Ⓔ
3 Ⓐ Ⓑ Ⓒ Ⓓ Ⓔ	28 Ⓐ Ⓑ Ⓒ Ⓓ Ⓔ	53 Ⓐ Ⓑ Ⓒ Ⓓ Ⓔ	78 Ⓐ Ⓑ Ⓒ Ⓓ Ⓔ
4 Ⓐ Ⓑ Ⓒ Ⓓ Ⓔ	29 Ⓐ Ⓑ Ⓒ Ⓓ Ⓔ	54 Ⓐ Ⓑ Ⓒ Ⓓ Ⓔ	79 Ⓐ Ⓑ Ⓒ Ⓓ Ⓔ
5 Ⓐ Ⓑ Ⓒ Ⓓ Ⓔ	30 Ⓐ Ⓑ Ⓒ Ⓓ Ⓔ	55 Ⓐ Ⓑ Ⓒ Ⓓ Ⓔ	80 Ⓐ Ⓑ Ⓒ Ⓓ Ⓔ
6 Ⓐ Ⓑ Ⓒ Ⓓ Ⓔ	31 Ⓐ Ⓑ Ⓒ Ⓓ Ⓔ	56 Ⓐ Ⓑ Ⓒ Ⓓ Ⓔ	81 Ⓐ Ⓑ Ⓒ Ⓓ Ⓔ
7 Ⓐ Ⓑ Ⓒ Ⓓ Ⓔ	32 Ⓐ Ⓑ Ⓒ Ⓓ Ⓔ	57 Ⓐ Ⓑ Ⓒ Ⓓ Ⓔ	82 Ⓐ Ⓑ Ⓒ Ⓓ Ⓔ
8 Ⓐ Ⓑ Ⓒ Ⓓ Ⓔ	33 Ⓐ Ⓑ Ⓒ Ⓓ Ⓔ	58 Ⓐ Ⓑ Ⓒ Ⓓ Ⓔ	83 Ⓐ Ⓑ Ⓒ Ⓓ Ⓔ
9 Ⓐ Ⓑ Ⓒ Ⓓ Ⓔ	34 Ⓐ Ⓑ Ⓒ Ⓓ Ⓔ	59 Ⓐ Ⓑ Ⓒ Ⓓ Ⓔ	84 Ⓐ Ⓑ Ⓒ Ⓓ Ⓔ
10 Ⓐ Ⓑ Ⓒ Ⓓ Ⓔ	35 Ⓐ Ⓑ Ⓒ Ⓓ Ⓔ	60 Ⓐ Ⓑ Ⓒ Ⓓ Ⓔ	85 Ⓐ Ⓑ Ⓒ Ⓓ Ⓔ
11 Ⓐ Ⓑ Ⓒ Ⓓ Ⓔ	36 Ⓐ Ⓑ Ⓒ Ⓓ Ⓔ	61 Ⓐ Ⓑ Ⓒ Ⓓ Ⓔ	86 Ⓐ Ⓑ Ⓒ Ⓓ Ⓔ
12 Ⓐ Ⓑ Ⓒ Ⓓ Ⓔ	37 Ⓐ Ⓑ Ⓒ Ⓓ Ⓔ	62 Ⓐ Ⓑ Ⓒ Ⓓ Ⓔ	87 Ⓐ Ⓑ Ⓒ Ⓓ Ⓔ
13 Ⓐ Ⓑ Ⓒ Ⓓ Ⓔ	38 Ⓐ Ⓑ Ⓒ Ⓓ Ⓔ	63 Ⓐ Ⓑ Ⓒ Ⓓ Ⓔ	88 Ⓐ Ⓑ Ⓒ Ⓓ Ⓔ
14 Ⓐ Ⓑ Ⓒ Ⓓ Ⓔ	39 Ⓐ Ⓑ Ⓒ Ⓓ Ⓔ	64 Ⓐ Ⓑ Ⓒ Ⓓ Ⓔ	89 Ⓐ Ⓑ Ⓒ Ⓓ Ⓔ
15 Ⓐ Ⓑ Ⓒ Ⓓ Ⓔ	40 Ⓐ Ⓑ Ⓒ Ⓓ Ⓔ	65 Ⓐ Ⓑ Ⓒ Ⓓ Ⓔ	90 Ⓐ Ⓑ Ⓒ Ⓓ Ⓔ
16 Ⓐ Ⓑ Ⓒ Ⓓ Ⓔ	41 Ⓐ Ⓑ Ⓒ Ⓓ Ⓔ	66 Ⓐ Ⓑ Ⓒ Ⓓ Ⓔ	91 Ⓐ Ⓑ Ⓒ Ⓓ Ⓔ
17 Ⓐ Ⓑ Ⓒ Ⓓ Ⓔ	42 Ⓐ Ⓑ Ⓒ Ⓓ Ⓔ	67 Ⓐ Ⓑ Ⓒ Ⓓ Ⓔ	92 Ⓐ Ⓑ Ⓒ Ⓓ Ⓔ
18 Ⓐ Ⓑ Ⓒ Ⓓ Ⓔ	43 Ⓐ Ⓑ Ⓒ Ⓓ Ⓔ	68 Ⓐ Ⓑ Ⓒ Ⓓ Ⓔ	93 Ⓐ Ⓑ Ⓒ Ⓓ Ⓔ
19 Ⓐ Ⓑ Ⓒ Ⓓ Ⓔ	44 Ⓐ Ⓑ Ⓒ Ⓓ Ⓔ	69 Ⓐ Ⓑ Ⓒ Ⓓ Ⓔ	94 Ⓐ Ⓑ Ⓒ Ⓓ Ⓔ
20 Ⓐ Ⓑ Ⓒ Ⓓ Ⓔ	45 Ⓐ Ⓑ Ⓒ Ⓓ Ⓔ	70 Ⓐ Ⓑ Ⓒ Ⓓ Ⓔ	95 Ⓐ Ⓑ Ⓒ Ⓓ Ⓔ
21 Ⓐ Ⓑ Ⓒ Ⓓ Ⓔ	46 Ⓐ Ⓑ Ⓒ Ⓓ Ⓔ	71 Ⓐ Ⓑ Ⓒ Ⓓ Ⓔ	96 Ⓐ Ⓑ Ⓒ Ⓓ Ⓔ
22 Ⓐ Ⓑ Ⓒ Ⓓ Ⓔ	47 Ⓐ Ⓑ Ⓒ Ⓓ Ⓔ	72 Ⓐ Ⓑ Ⓒ Ⓓ Ⓔ	97 Ⓐ Ⓑ Ⓒ Ⓓ Ⓔ
23 Ⓐ Ⓑ Ⓒ Ⓓ Ⓔ	48 Ⓐ Ⓑ Ⓒ Ⓓ Ⓔ	73 Ⓐ Ⓑ Ⓒ Ⓓ Ⓔ	98 Ⓐ Ⓑ Ⓒ Ⓓ Ⓔ
24 Ⓐ Ⓑ Ⓒ Ⓓ Ⓔ	49 Ⓐ Ⓑ Ⓒ Ⓓ Ⓔ	74 Ⓐ Ⓑ Ⓒ Ⓓ Ⓔ	99 Ⓐ Ⓑ Ⓒ Ⓓ Ⓔ
25 Ⓐ Ⓑ Ⓒ Ⓓ Ⓔ	50 Ⓐ Ⓑ Ⓒ Ⓓ Ⓔ	75 Ⓐ Ⓑ Ⓒ Ⓓ Ⓔ	100 Ⓐ Ⓑ Ⓒ Ⓓ Ⓔ

Diagnostic Test Form ○ Side 1

1. YOUR NAME: _____
(Print)   Last   First   M.I.

SIGNATURE: _____ **DATE:** _____ / _____ / _____

HOME ADDRESS: _____
(Print)   Number and Street

          E-MAIL: _____

City   State   Zip

PHONE NO.: _____ **SCHOOL:** _____ **CLASS OF:** _____
(Print)

IMPORTANT: Please fill in these boxes exactly as shown on the back cover of your test book.

OpScan *i*NSIGHT™ forms by Pearson NCS EM-255325-1:654321
Printed in U.S.A.

© TPR Education IP Holdings, LLC

2. TEST FORM

_____

5. YOUR NAME

First 4 letters of last name				FIRST INIT	MID INIT
Ⓐ	Ⓐ	Ⓐ	Ⓐ	Ⓐ	Ⓐ
Ⓑ	Ⓑ	Ⓑ	Ⓑ	Ⓑ	Ⓑ
Ⓒ	Ⓒ	Ⓒ	Ⓒ	Ⓒ	Ⓒ
Ⓓ	Ⓓ	Ⓓ	Ⓓ	Ⓓ	Ⓓ
Ⓔ	Ⓔ	Ⓔ	Ⓔ	Ⓔ	Ⓔ
Ⓕ	Ⓕ	Ⓕ	Ⓕ	Ⓕ	Ⓕ
Ⓖ	Ⓖ	Ⓖ	Ⓖ	Ⓖ	Ⓖ
Ⓗ	Ⓗ	Ⓗ	Ⓗ	Ⓗ	Ⓗ
Ⓘ	Ⓘ	Ⓘ	Ⓘ	Ⓘ	Ⓘ
Ⓙ	Ⓙ	Ⓙ	Ⓙ	Ⓙ	Ⓙ
Ⓚ	Ⓚ	Ⓚ	Ⓚ	Ⓚ	Ⓚ
Ⓛ	Ⓛ	Ⓛ	Ⓛ	Ⓛ	Ⓛ
Ⓜ	Ⓜ	Ⓜ	Ⓜ	Ⓜ	Ⓜ
Ⓝ	Ⓝ	Ⓝ	Ⓝ	Ⓝ	Ⓝ
Ⓞ	Ⓞ	Ⓞ	Ⓞ	Ⓞ	Ⓞ
Ⓟ	Ⓟ	Ⓟ	Ⓟ	Ⓟ	Ⓟ
Ⓠ	Ⓠ	Ⓠ	Ⓠ	Ⓠ	Ⓠ
Ⓡ	Ⓡ	Ⓡ	Ⓡ	Ⓡ	Ⓡ
Ⓢ	Ⓢ	Ⓢ	Ⓢ	Ⓢ	Ⓢ
Ⓣ	Ⓣ	Ⓣ	Ⓣ	Ⓣ	Ⓣ
Ⓤ	Ⓤ	Ⓤ	Ⓤ	Ⓤ	Ⓤ
Ⓥ	Ⓥ	Ⓥ	Ⓥ	Ⓥ	Ⓥ
Ⓦ	Ⓦ	Ⓦ	Ⓦ	Ⓦ	Ⓦ
Ⓧ	Ⓧ	Ⓧ	Ⓧ	Ⓧ	Ⓧ
Ⓨ	Ⓨ	Ⓨ	Ⓨ	Ⓨ	Ⓨ
Ⓩ	Ⓩ	Ⓩ	Ⓩ	Ⓩ	Ⓩ

3. TEST CODE   4. PHONE NUMBER

(bubbles 0–9 across all columns)

⓪ ⓪ ⓪ ⓪	⓪ ⓪ ⓪ ⓪ ⓪ ⓪ ⓪
① ① ① ①	① ① ① ① ① ① ①
② ② ② ②	② ② ② ② ② ② ②
③ ③ ③ ③	③ ③ ③ ③ ③ ③ ③
④ ④ ④ ④	④ ④ ④ ④ ④ ④ ④
⑤ ⑤ ⑤ ⑤	⑤ ⑤ ⑤ ⑤ ⑤ ⑤ ⑤
⑥ ⑥ ⑥ ⑥	⑥ ⑥ ⑥ ⑥ ⑥ ⑥ ⑥
⑦ ⑦ ⑦ ⑦	⑦ ⑦ ⑦ ⑦ ⑦ ⑦ ⑦
⑧ ⑧ ⑧ ⑧	⑧ ⑧ ⑧ ⑧ ⑧ ⑧ ⑧
⑨ ⑨ ⑨ ⑨	⑨ ⑨ ⑨ ⑨ ⑨ ⑨ ⑨

6. DATE OF BIRTH

MONTH	DAY		YEAR	
○ JAN				
○ FEB				
○ MAR	⓪	⓪	⓪	⓪
○ APR	①	①	①	①
○ MAY	②	②	②	②
○ JUN	③	③	③	③
○ JUL		④	④	④
○ AUG		⑤	⑤	⑤
○ SEP		⑥	⑥	⑥
○ OCT		⑦	⑦	⑦
○ NOV		⑧	⑧	⑧
○ DEC		⑨	⑨	⑨

7. SEX
○ MALE
○ FEMALE

8. OTHER
1 Ⓐ Ⓑ Ⓒ Ⓓ Ⓔ
2 Ⓐ Ⓑ Ⓒ Ⓓ Ⓔ
3 Ⓐ Ⓑ Ⓒ Ⓓ Ⓔ

Begin with number 1 for each new section of the test. Leave blank any extra answer spaces.

SECTION 1

1 Ⓐ Ⓑ Ⓒ Ⓓ Ⓔ	26 Ⓐ Ⓑ Ⓒ Ⓓ Ⓔ	51 Ⓐ Ⓑ Ⓒ Ⓓ Ⓔ	76 Ⓐ Ⓑ Ⓒ Ⓓ Ⓔ
2 Ⓐ Ⓑ Ⓒ Ⓓ Ⓔ	27 Ⓐ Ⓑ Ⓒ Ⓓ Ⓔ	52 Ⓐ Ⓑ Ⓒ Ⓓ Ⓔ	77 Ⓐ Ⓑ Ⓒ Ⓓ Ⓔ
3 Ⓐ Ⓑ Ⓒ Ⓓ Ⓔ	28 Ⓐ Ⓑ Ⓒ Ⓓ Ⓔ	53 Ⓐ Ⓑ Ⓒ Ⓓ Ⓔ	78 Ⓐ Ⓑ Ⓒ Ⓓ Ⓔ
4 Ⓐ Ⓑ Ⓒ Ⓓ Ⓔ	29 Ⓐ Ⓑ Ⓒ Ⓓ Ⓔ	54 Ⓐ Ⓑ Ⓒ Ⓓ Ⓔ	79 Ⓐ Ⓑ Ⓒ Ⓓ Ⓔ
5 Ⓐ Ⓑ Ⓒ Ⓓ Ⓔ	30 Ⓐ Ⓑ Ⓒ Ⓓ Ⓔ	55 Ⓐ Ⓑ Ⓒ Ⓓ Ⓔ	80 Ⓐ Ⓑ Ⓒ Ⓓ Ⓔ
6 Ⓐ Ⓑ Ⓒ Ⓓ Ⓔ	31 Ⓐ Ⓑ Ⓒ Ⓓ Ⓔ	56 Ⓐ Ⓑ Ⓒ Ⓓ Ⓔ	81 Ⓐ Ⓑ Ⓒ Ⓓ Ⓔ
7 Ⓐ Ⓑ Ⓒ Ⓓ Ⓔ	32 Ⓐ Ⓑ Ⓒ Ⓓ Ⓔ	57 Ⓐ Ⓑ Ⓒ Ⓓ Ⓔ	82 Ⓐ Ⓑ Ⓒ Ⓓ Ⓔ
8 Ⓐ Ⓑ Ⓒ Ⓓ Ⓔ	33 Ⓐ Ⓑ Ⓒ Ⓓ Ⓔ	58 Ⓐ Ⓑ Ⓒ Ⓓ Ⓔ	83 Ⓐ Ⓑ Ⓒ Ⓓ Ⓔ
9 Ⓐ Ⓑ Ⓒ Ⓓ Ⓔ	34 Ⓐ Ⓑ Ⓒ Ⓓ Ⓔ	59 Ⓐ Ⓑ Ⓒ Ⓓ Ⓔ	84 Ⓐ Ⓑ Ⓒ Ⓓ Ⓔ
10 Ⓐ Ⓑ Ⓒ Ⓓ Ⓔ	35 Ⓐ Ⓑ Ⓒ Ⓓ Ⓔ	60 Ⓐ Ⓑ Ⓒ Ⓓ Ⓔ	85 Ⓐ Ⓑ Ⓒ Ⓓ Ⓔ
11 Ⓐ Ⓑ Ⓒ Ⓓ Ⓔ	36 Ⓐ Ⓑ Ⓒ Ⓓ Ⓔ	61 Ⓐ Ⓑ Ⓒ Ⓓ Ⓔ	86 Ⓐ Ⓑ Ⓒ Ⓓ Ⓔ
12 Ⓐ Ⓑ Ⓒ Ⓓ Ⓔ	37 Ⓐ Ⓑ Ⓒ Ⓓ Ⓔ	62 Ⓐ Ⓑ Ⓒ Ⓓ Ⓔ	87 Ⓐ Ⓑ Ⓒ Ⓓ Ⓔ
13 Ⓐ Ⓑ Ⓒ Ⓓ Ⓔ	38 Ⓐ Ⓑ Ⓒ Ⓓ Ⓔ	63 Ⓐ Ⓑ Ⓒ Ⓓ Ⓔ	88 Ⓐ Ⓑ Ⓒ Ⓓ Ⓔ
14 Ⓐ Ⓑ Ⓒ Ⓓ Ⓔ	39 Ⓐ Ⓑ Ⓒ Ⓓ Ⓔ	64 Ⓐ Ⓑ Ⓒ Ⓓ Ⓔ	89 Ⓐ Ⓑ Ⓒ Ⓓ Ⓔ
15 Ⓐ Ⓑ Ⓒ Ⓓ Ⓔ	40 Ⓐ Ⓑ Ⓒ Ⓓ Ⓔ	65 Ⓐ Ⓑ Ⓒ Ⓓ Ⓔ	90 Ⓐ Ⓑ Ⓒ Ⓓ Ⓔ
16 Ⓐ Ⓑ Ⓒ Ⓓ Ⓔ	41 Ⓐ Ⓑ Ⓒ Ⓓ Ⓔ	66 Ⓐ Ⓑ Ⓒ Ⓓ Ⓔ	91 Ⓐ Ⓑ Ⓒ Ⓓ Ⓔ
17 Ⓐ Ⓑ Ⓒ Ⓓ Ⓔ	42 Ⓐ Ⓑ Ⓒ Ⓓ Ⓔ	67 Ⓐ Ⓑ Ⓒ Ⓓ Ⓔ	92 Ⓐ Ⓑ Ⓒ Ⓓ Ⓔ
18 Ⓐ Ⓑ Ⓒ Ⓓ Ⓔ	43 Ⓐ Ⓑ Ⓒ Ⓓ Ⓔ	68 Ⓐ Ⓑ Ⓒ Ⓓ Ⓔ	93 Ⓐ Ⓑ Ⓒ Ⓓ Ⓔ
19 Ⓐ Ⓑ Ⓒ Ⓓ Ⓔ	44 Ⓐ Ⓑ Ⓒ Ⓓ Ⓔ	69 Ⓐ Ⓑ Ⓒ Ⓓ Ⓔ	94 Ⓐ Ⓑ Ⓒ Ⓓ Ⓔ
20 Ⓐ Ⓑ Ⓒ Ⓓ Ⓔ	45 Ⓐ Ⓑ Ⓒ Ⓓ Ⓔ	70 Ⓐ Ⓑ Ⓒ Ⓓ Ⓔ	95 Ⓐ Ⓑ Ⓒ Ⓓ Ⓔ
21 Ⓐ Ⓑ Ⓒ Ⓓ Ⓔ	46 Ⓐ Ⓑ Ⓒ Ⓓ Ⓔ	71 Ⓐ Ⓑ Ⓒ Ⓓ Ⓔ	96 Ⓐ Ⓑ Ⓒ Ⓓ Ⓔ
22 Ⓐ Ⓑ Ⓒ Ⓓ Ⓔ	47 Ⓐ Ⓑ Ⓒ Ⓓ Ⓔ	72 Ⓐ Ⓑ Ⓒ Ⓓ Ⓔ	97 Ⓐ Ⓑ Ⓒ Ⓓ Ⓔ
23 Ⓐ Ⓑ Ⓒ Ⓓ Ⓔ	48 Ⓐ Ⓑ Ⓒ Ⓓ Ⓔ	73 Ⓐ Ⓑ Ⓒ Ⓓ Ⓔ	98 Ⓐ Ⓑ Ⓒ Ⓓ Ⓔ
24 Ⓐ Ⓑ Ⓒ Ⓓ Ⓔ	49 Ⓐ Ⓑ Ⓒ Ⓓ Ⓔ	74 Ⓐ Ⓑ Ⓒ Ⓓ Ⓔ	99 Ⓐ Ⓑ Ⓒ Ⓓ Ⓔ
25 Ⓐ Ⓑ Ⓒ Ⓓ Ⓔ	50 Ⓐ Ⓑ Ⓒ Ⓓ Ⓔ	75 Ⓐ Ⓑ Ⓒ Ⓓ Ⓔ	100 Ⓐ Ⓑ Ⓒ Ⓓ Ⓔ

NOTES

NOTES

NOTES

NOTES

NOTES

NOTES

NOTES

NOTES

NOTES

NOTES

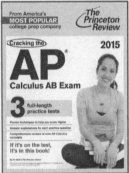

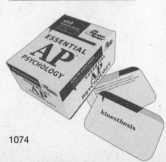